Electronic
Digital System Fundamentals

Electronic
Digital System Fundamentals

Dale Patrick
Stephen Fardo
Vigyan 'Vigs' Chandra

THE FAIRMONT PRESS, INC.

CRC Press
Taylor & Francis Group

Library of Congress Cataloging-in-Publication Data

Patrick, Dale R.
    Electronic digital system fundamentals / Dale Patrick, Stephen Fardo,
Vigyan 'Vigs' Chandra.
        p. cm.
        Includes index.   1005508493
        ISBN 0-88173-540-X (alk. paper) -- ISBN 0-88173-541-8 (electronic) -- ISBN
1-4200-6774-5 (Taylor & Francis distribution : alk. paper)
        1. Digital electronics. I. Fardo, Stephen W. II. Chandra, Vigyan, 1968-
III. Title.

    TK7868.D5P378 2008
    621.381--dc22

                                                                    2007032778

Published by The Fairmont Press, Inc.
700 Indian Trail
Lilburn, GA 30047
tel: 770-925-9388; fax: 770-381-9865
http://www.fairmontpress.com

Distributed by Taylor & Francis Ltd.
6000 Broken Sound Parkway NW, Suite 300
Boca Raton, FL 33487, USA
E-mail: orders@crcpress.com

Distributed by Taylor & Francis Ltd.
23-25 Blades Court
Deodar Road
London SW15 2NU, UK
E-mail: uk.tandf@thomsonpublishingservices.co.uk

Printed in the United States of America
10 9 8 7 6 5 4 3 2 1

0-88173-540-X (The Fairmont Press, Inc.)
1-4200-6774-5 (Taylor & Francis Ltd.)

iv

Table of Contents

Preface

Electronic Digital Systems Fundamentals is an introductory text that provides coverage of the various topics in the field of digital electronics. The key concepts presented in this book are discussed using a simplified approach that greatly enhances learning. The use of mathematics is kept to the very minimum and is discussed clearly through applications and illustrations.

Each chapter is organized in a step-by-step progression of concepts and theory. The chapters begin with an introduction, discuss important concepts with the help of numerous illustrations, as well as examples, and conclude with summaries.

The overall learning objectives of this book include:

- Describe the characteristics of a digital electronic system.
- Explain the operation of digital electronic gate circuits.
- Demonstrate how gate functions are achieved.
- Use binary, octal, and hexadecimal counting systems.
- Use Boolean algebra to define different logic operations.
- Change a logic diagram into a Boolean expression and a Boolean expression into a logic diagram.
- Explain how discrete components are utilized in the construction of digital integrated circuits.
- Discuss how counting, decoding, multiplexing, demultiplexing, and clocks function with logic devices.
- Change a truth table into a logic expression and a logic expression into a truth table.
- Identify some of the common functions of digital memory.
- Explain how arithmetic operations are achieved with digital circuitry.

Appendices are also included that contain information regarding circuit symbols, data sheets and electrical safety.

The authors hope that you will find Electronic Digital System Fundamentals easy to understand and that you are successful in your pursuit of knowledge in this exciting technical area.

Dale R. Patrick,
Stephen W. Fardo,
Vigyan 'Vigs' Chandra
Richmond, Kentucky

Chapter 1

Introduction to Digital Systems

Chapter 1 provides an overview of electronic digital systems. The concepts discussed in this chapter are important for developing an understanding of electronic digital systems. Digital electronics is undoubtedly the fastest growing area in the field of electronics today. Personal computers, cameras, cell phones, calculators, watches, clocks, video games, test instruments and home appliances are only a few of the applications of digital systems. Digital systems play an essential role in our daily lives and new applications are emerging at a rapid pace.

DIGITAL AND ANALOG ELECTRONICS SYSTEMS

Electronics is further divided into two main categories: analog and digital. Analog electronics deals with the analog systems, in which signals are free to take any possible numerical value. Digital electronics deals with digital or discrete systems, which has signals that take on only a limited range of values. Practical systems are often hybrids having both analog and discrete components.

Analog as in the term 'analogous', is used to represent the variation of an electrical quantity when a corresponding physical phenomenon varies. For example, when the flow of fluid through a pipe increases, an analog meter monitoring the flow may generate a larger voltage (or other electric quantity), which can then be displayed on a scale calibrated to indicate flow rate. Most quantities in nature are inherently analog—temperature, pressure, flow, light intensity change, loudness of sound, current flow in a circuit, or voltage variations.

Digital signals are characterized by discrete variations or jumps in their values. They are useful in producing information about a system. For example, in the case of a sensor monitoring the flow rate in a water canal, it might be sufficient to know whether the flow has reached a critical level, rather than monitoring every possible value of the flow. All values below this critical flow value could be regarded as part of the normal functioning of the system. Hence, when the critical flow value is passed the sensor could trip (switch on), and for normal flow values it would remain off. It can be seen right away that only the values of interest are being used (non-critical flow, critical flow). These in turn can be represented by two conditions of a flow switch—open when the flow is non-critical, and closed when the flow has reached critical.

Figures 1-1(a) and (b) show two conditions of fluid flow through a water pipe, and the corresponding digital flow switch conditions measured by a sensor. Compare it with the graph given for the real-time analog fluid flow rate in the pipe given in (c).

If the switch is connected to a voltage source, then with the flow switch open, no voltage would appear across the buzzer, and the voltage would be 0V. On the other hand when the flow switch is closed, the supply voltage (5V) would appear on the other side of the buzzer. Any digital system receiving a 5V signal would know right away that the flow has reached critical level. Otherwise the system is functioning at a non-critical level (normal flow or even no flow). The process of digitizing the analog signal is shown in Figure 1-2. This might require scaling of the voltage received from the sensor before being applied to a digital circuit. This is because digital circuits require voltage in certain range, 0-5V, before they can

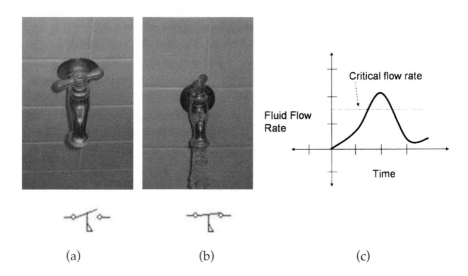

(a)   (b)   (c)

Figure 1-1. Monitoring fluid-flow in a pipe

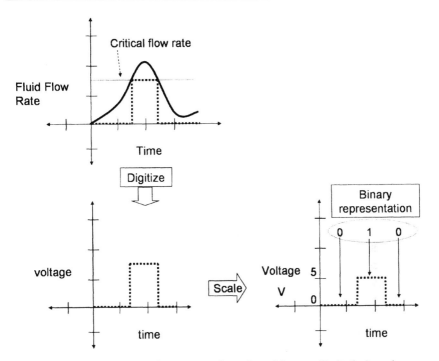

Figure 1-2. Converting an analog signal into a digital signal

function properly.

Digital electronics is considered to be a counting operation. A digital watch tells time by counting generated pulses. The resulting count is then displayed by numbers representing hours, minutes, and seconds. A computer also has an electronic clock that generates pulses. These pulses are counted and in many cases manipulated to perform a control function. Digital circuits can store signal data, retrieve them when needed, and make operational decisions.

ADVANTAGES OF DIGITAL SYSTEMS

- Storage space in digital devices can be increased or decreased based on the application. While hard disks used inside computer systems can store enormous quantities of data in various electronic formats, other mobile devices such as cell phones are limited in their storage.

- The accuracy of digital devices can also be increased based on the precision needed in an application.

- Digital devices are less susceptible to electrical interference, temperature and humidity variations as compared to analog devices, since they uses discrete values corresponding to different values, not a continuous range of values.

- Digital devices can be mass manufactured, and with the increase in fabrication technologies, the number of defects in manufactured integrated circuits (ICs) has reduced considerably.

- The design of digital systems is easier as compared to analog systems. This is in part because progressively larger digital systems can be built using the same principles which apply to much smaller digital systems.

- There are several different types of programmable digital devices. This makes it possible to change the functionality of a device.

DISADVANTAGES OF DIGITAL SYSTEMS

- The world around us is analog in general. For example it has continuous variations in temperature, pressure, flow, pressure, sound and light intensity. For a digital system to process this type of information, some accuracy will be sacrificed and delays due to conversion and processing times will be introduced.

- Digital devices use components such as transistors which exhibit analog behavior and it is important to ensure that theses properties do not dominate in the digital circuit.

DIGITAL SYSTEM OPERATIONAL STATES

Digital systems require a precise definition of operational states or conditions in order to be useful. In practice, binary signals can be processed very easily through electronic circuitry because they can be represented by two stable states of operation. These states can be easily defined as on or off, 1 or 0, up or down, voltage or no voltage, right or left, or any of the other two-condition designations. There must be no in-between step or condition. These states must be decidedly different and easily distinguished.

The symbols used to define the operational state of a binary system are very important. In positive binary logic, such things as voltage, on, true, or a letter designation such as 'A' are used to denote the 1 operational state. No voltage, off, false, or the letter $\bar{A}$ are commonly used to denote the alternate, or 0, condition. An operating system can be set to either state, where it will remain until something causes it to change conditions.

Any device that can be set in one of two operational states or conditions by an outside signal is said to be bistable. Switches, relays, transistors, diodes, and ICs are commonly used examples. In a strict sense, a bistable device has the capability of storing one binary digit or bit of information. By employing a number of these devices, it is possible to build an electronic circuit that will make decisions based on the applied input signals. The output of such a circuit is, therefore, a decision based on the operational conditions of the input. Since this application of a bistable device makes logical decisions, it is commonly called a binary logic circuit, or simply a logic circuit.

There are two basic types of logic circuits in a digital system. One type of logic circuit is designed to make decisions. It has data applied to its input and produces an output that coincides with a prescribed combination of rules. Electronic decisions are made with logic gates. Memory is the other type of logic circuit. Memory circuits store binary data. These data can be stored and retrieved from memory when the need arises. Special ICs are used to achieve the memory function of a digital system. Memory is a primary function of a digital system. Performance is largely dependent on the capacity of a system's memory.

BINARY LOGIC LEVELS

The term 'binary' is derived from the term 'bi' meaning two. A binary number system thus has two numbers, and since all non-negative numbers in any number system begin at '0', this is the first number. The second number is '1'.

Almost all modern day computer systems and electronic devices use circuits which accept inputs which can have exactly two states. These devices process information and generate outputs each of which can have exactly two states as well. The two states correspond to two voltage ranges or levels designated as 'low' and 'high'.

Electronic devices normally accept inputs which are in the interval 0V-5V. Some part of this interval is designated as the low level, and another as the high level. In order to ensure that these two ranges do not overlap, they are separated by an intermediate range. This is shown in Figure 1-3.

Since digital devices operate in either the low range or the high range of voltage, it is important that while switching between these levels, the transition be as quick as possible, minimizing the time spent in the intermediate range. The reason is that the behavior of digital devices is unpredictable when their inputs are not in the valid low or high ranges.

BINARY NUMBER SYSTEM

The binary number system, with its use of two numerals, 0 and 1, are referred to as 'low' and 'high' levels, finds numerous applications in digital circuits. As with the decimal number system more than one digit may be used for expressing larger quantities.

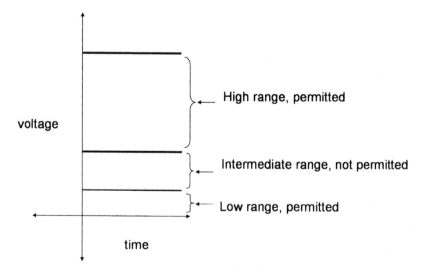

Figure 1-3. Voltage ranges for Low and High sensed by digital devices

BIT

Each binary digit is abbreviated as a 'bit' ('bi' from binary, and the 't' from the digit). Each bit can take on 2 values, 0 or 1. This is shown in Figure 1-4.

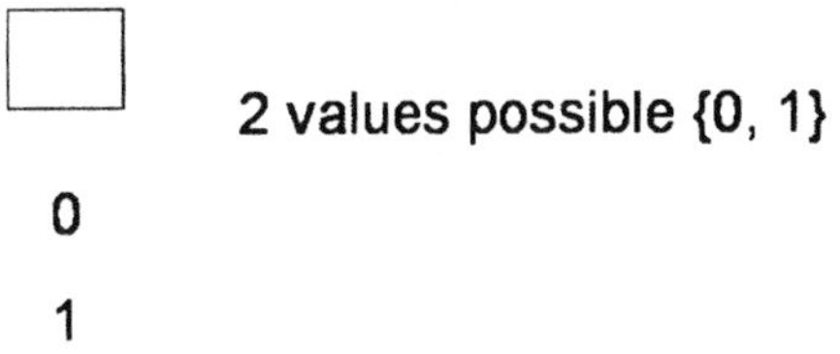

2 values possible {0, 1}

0

1

Figure 1-4. Enumerating all possible single-bit values

The bit is used most often for expressing the status of a digital input or output. For example the input of a push-button switch to a digital system may cause a 0V or a 5V to be applied or removed based on the switch connections. Similarly, the output of a digital system driving a buzzer for example, may be at 0V (off) or 5V (on).

When more than one bit it used it can be used to represent larger quantities. With 2 bits for example, each bit is permitted to take on $2 = 2^1 = 2$, with the values 0 or 1, there are a total of $2 \times 2 = 2^2 = 4$ possible values that can be taken. This is shown in Figure 1-5.

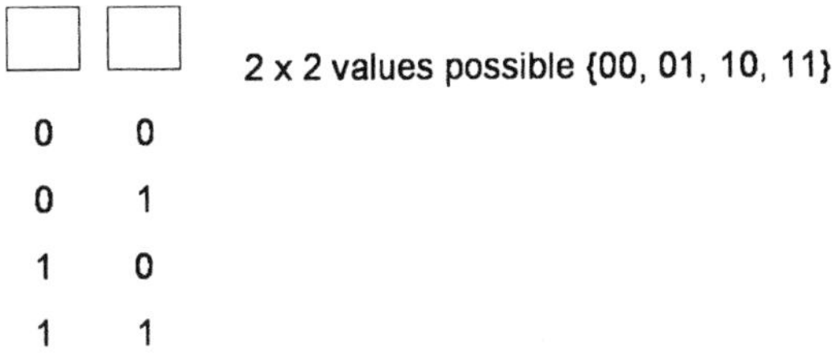

2 x 2 values possible {00, 01, 10, 11}

0   0

0   1

1   0

1   1

Figure 1-5. Enumerating all possible 2-bit values

DISCRETE AND INTEGRATED CIRCUITS

Discrete circuits are created when electrical components such as resistors, capacitors; and transistors are manufactured separately then connected together forming a circuit either using wires or conducting tracks on printed circuit boards. Discrete circuits take up considerable space and generate heat. They also require wiring with soldered contacts for joining the different circuit components together. For reasonably large size circuits with hundreds of components such as those used in washing machines or VCRs, the need for external wiring including subsequent soldered points creates reliability issues. Miniaturization of circuit components solves some of these issues but the need for external wiring and soldering still exists, and at high frequencies as are present in computers these act as tiny antennas. This causes interference, wherein the signal radiated out by a component or on wires can be picked up by others.

Integrated circuits (ICs) are monolithic device which would incorporate electrical components such as resistors, capacitors, and semi-conductors such as transistors, diodes, are interconnected in a single package. In discrete components there is a need to connect all devices together for creating larger circuits, with the connections being soldered. Handling the hundreds and thousands of components and their associated wiring came to be termed as the 'tyranny of numbers'. The solution to this was first proposed by Robert Noyce and Jack Kirby at approximately the same time and independent of each other. This was done at a time when miniaturization of discrete components was nearing its physical limits and wiring between these minute components was becoming increasing hard to manufacture. ICs made it possible to create all these devices on a slice of semiconductor material, whose electrical conductivity can be manipulated. Owing to mass manufacturing techniques the reliability of ICs is phenomenal. Several million of these devices can be manufactured simultaneously, as in the case of modern day microprocessors, out of the same piece of semi-conductor material such as silicon or germanium. They weigh considerably less, and take up less space, and generate less heat, thus consuming less power. However, if any sub-component of an IC fails the entire device needs to be replaced. Once manufactured the properties of all circuit components is set and cannot be altered. Additionally, only very small capacitors can be manufactured, and in the past it has not been possible to manufacture inductors and transformers on an IC.

Over the years as the complexity of digital devices has expanded phenomenally, the space requirement has shrunk at the same rate. This phenomenon observed by Gordon Moore, which bears the important law after his name stating that the number of electronic devices (transistors) and resistors used on a chip doubles every 18 months.

Semiconductors as the name suggests do not function quite like conductors at room temperature. In fact, they have little or no conductivity at room temperature. However, by a process of doping (adding minute quantities of other materials) the conductivity of the material can be substantially enhanced. When the doping process produces an excess of electrons in the semiconductor a 'n' material is said to have been created. On the other hand, when the doping process produces a material with a deficit of electrons in the semiconductor, a 'p' material is said to have been created.

When p and n materials are joined to each other, the structure is called the 'pn junction'. At this junction there is an initial diffusion of excess electrons from the n into the p region which has a deficit of electrons. After awhile the diffusion process stops. The portion of the n region at the junction which lost electrons gains a net positive charge, whereas the portion of the p region at the junction which gains electrons gains a net negative charge. Overall the junction thus develops a minute potential, approximately 0.7V for silicon semiconductors and 0.3V for germanium semiconductors. This is shown Figure 1-6. The symbol and crystal structure of the diode is shown in Figure 1-6 (a), and the photograph of a diode is shown in Figure 1-6 (b). In a diode the p region is designated as the anode and the n region as the cathode.

The resultant semiconductor component is called a 'diode', which permits current flow in one direction but blocks it in the opposite direction. It can thus be used as a switch. Since it is fabricated using a block of semiconductor material by varying the doping of different regions, it is also a type of integrated circuit or IC. ICs are generally referred to as a 'chip'. When volt-

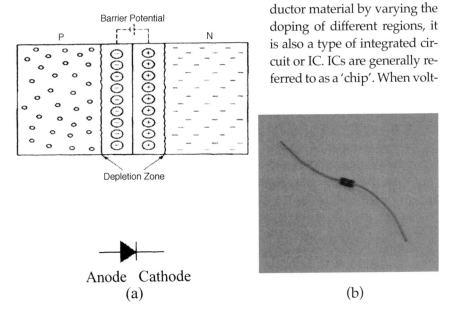

Anode   Cathode
(a)   (b)

Figure 1-6. pn junction

age is applied across the diode such that the anode is connected to the positive and the cathode the negative, electrons can flow across the junction, and current flow established. As the electrons move from the n region to the p region they lose energy, which is dissipated usually in the form of heat. In the case of light emitting diodes or LEDs this energy is dissipated in the form of light as shown in Figure 1-7. A current limiting resistor, usually between 200-1000 Ω should be used in a LED circuit, when used with voltage source (3-6 V). This restricts the current flow to be well within safe operating values for the LED.

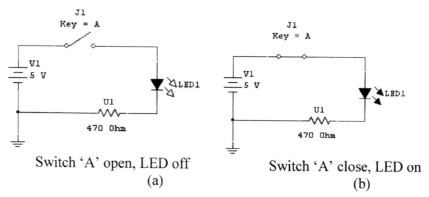

Switch 'A' open, LED off   Switch 'A' close, LED on
(a)   (b)

Figure 1-7. Operation of an LED

It is possible to create other electronic components on ICs such as:
• resistors (a heavily doped semiconductor material),
• transistors (a device with 2 pn junctions suitably arranged),
• capacitors (a p and n material separated by an insulator such as silicon di-oxide.

ICs are now used in almost every electronic device today—ranging from calculators, microprocessors, digital phones, personal computers, and cameras. Simple ICs perform specific functions and have a fewer number of pins, whereas complex ICs may offer programming, storage functions and have many pins.

Digital devices are built using two predominant types of technologies—TTL (Transistor Transistor Logic) and CMOS (Complementary Metal Oxide Semiconductor). Both of these have specific ranges for low and high for both the input and the outputs they generate. These transistor technologies themselves are described next.

TTL (TRANSISTOR-TRANSISTOR LOGIC)

TTL gates were first introduced by Texas Instruments or TI in the early 1960s. These devices were meant for educational, industry and experimental use. They were and still are marketed under the 74 _ _ series designation, where the '_ _' were decimal numerals that specify the particular type of operation being performed.

74XX is the common abbreviation used when referring to these devices, where the Xs stand for decimal numerals. It is common practice also to drop the 74 portion and refer to the gate simply by the latter portion which specifies the type of function the device has. For example the 7408 chip implements the AND logic function. It is commonly referred to as the 08 chip. The 74XX series is designed to operate in the temperature range of 0°C to 75°C. TI also manufactured the 54XX series which were meant primarily for military applications and could operate in the range -55°C to 125°C.

Over the years there have been improvements in the speed and power requirements of the TTL devices. With the use of a special type of transistor called the Schottky transistor, there was a great improvement in speed with these devices being called the 74SXX series, with the 'S' standing for the Schottky devices. However, there was an increase in power requirements, and a sub-family called the 74LSXX with the 'L' standing for low- power device. In this text most of the devices used will be of the 74LSXX series, for example the 74LS08 would then be a Low-power Schottky family quad AND gate. Further improvements led to the 74ASXX where the 'A' signifies advanced and the 74ALSXX, where the ALS signifies the advanced low-power Schottky devices.

Usually more than one gate is fabricated on an IC. In Figure 1-8(a) the gate level layout of the 7408 chip is shown, which implements the AND function. It has 4 AND logic gates packaged into one chip, thus being termed as a quad 2-input AND gate. Each AND gate uses 3 pins (2 for input and 1 for the output). The 4 gates of a 7408 needs 3 x 4 = 12 pins. For its operation the chip also needs a source of power, which is designated as VCC and ground, or GND. Thus, the total number of pins this chip has: 12 + 2 = 14. The 7408 IC is packaged as a Dual Inline Pin, or DIP package, and the pins arranged in two rows. This is shown in Figure 1-8(b).

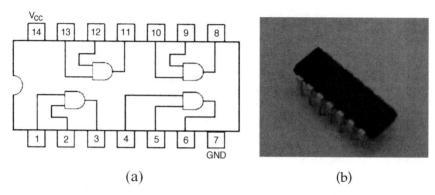

(a)   (b)

Figure 1-8. TTL 'AND' gate

CASCADING TTL DIGITAL DEVICES

Electronic devices are often used in conjunction with other devices and it is important that the output of one device produces a voltage which is within the acceptable range of logic low and logic high voltages. This is shown in Figure 1-9.

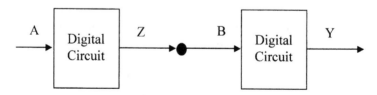

Figure 1-9 Cascading two digital circuits

Figure 1-10 shows the acceptable range of input voltages that may be applied, and the range of output voltages produced.

Note that the maximum output voltage of 'Z' corresponding to a low level (0.4V) is well within the range of the acceptable low input voltage of 'B' (0.8V); and the minimum output voltage of 'Z' corresponding the a high level (2.4V) is also well within the range of the acceptable logic high input voltage (2V). This ensures that the output of one system can be safely applied to the input of a subsequent digital device.

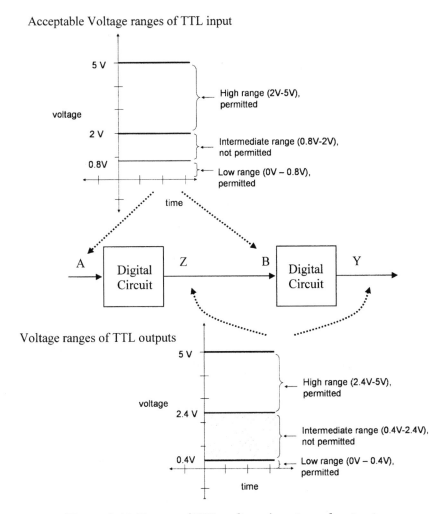

Figure 1-10. Range of TTL voltage inputs and outputs

CMOS (COMPLEMENTARY METAL OXIDE SEMICONDUCTOR)

MOSFETs were first introduced by in the late 1960s. They were marketed first under the 4000 series designation. However, in order to be competitive with TTL devices which offered similar functionality, a new series of CMOS chips was developed which used the 74C _ _, with the 'C' standing for CMOS and the '_ _' as before representing the particular functionality being implemented. For example, the AND gate equivalent to the 74LS08 is the 74C08. As with TTL devices, further upgrades in the technology made it possible to operate the CMOS devices faster, and were given the designation 74HC _ _, where the 'H' stands for high-speed. 74HCXX is the common abbreviation used when referring to these devices, where the Xs stand for decimal numerals. Further additions led to the development of 74ACXX (with the 'A' standing for 'Advanced') and 74ACTXX (with the 'ACT' standing for 'Advanced CMOS TTL compatible') devices.

IC FAMILIES—SCALE OF INTEGRATION (NUMBER OF GATES)

The number of transistors and gates being fabricated using these elements on integrated circuits has increased tremendously over the past few decades. Earlier ICs used small scale integration or SSI, in which the number of transistors were in the tens. Present day ultra large scale integration or ULSI techniques fabricate over a million transistors on an IC. The Pentium 4 extreme edition processor for example, has over 175 million transistors. Most computer systems are regarded as VLSI systems, rather than the more specific ULSI. The abbreviations associated with the scale of integration are shown in Figure 1-11.

Abbreviation	Scale of Integration	No. of Transistors (approximate)
SSI	Small Scale Integration	10
MSI	Medium Scale Integration	100
LSI	Large Scale Integration	1,000
VLSI	Very Large Scale Integration	10,000
SLSI	Super Large Scale Integration	100,000
ULSI	Ultra Large Scale Integration	1,000,000

Figure 1-11. Scale of integration used by digital devices

MONITORING AND CONTROL OF COMPLEX
SYSTEMS USING DIGITAL CIRCUITS

Almost all man-made systems such as those used in manufacturing, the military, communication networks, or computers require close monitoring and control. This is true also of several naturally occurring systems such as the weather conditions, forest fires, or fish populations. A system requiring this form of monitoring or control is shown in Figure 1-12. It could be one which monitors or controls the temperature in a furnace, the timing of an internal combustion engine, the countdown system for a missile launch system, road traffic light control, or data routing in communication systems.

Figure 1-12. Systems view

In one of its simplest implementations, the operation of an industrial system can be monitored by feedback signal(s) such as those generated by sensors. This is shown in Figure 1-13. The rectangular block houses the digital monitoring or control circuitry, and is called a 'block diagram'. It hides the actual implementation details and is helpful in understanding the higher level ideas about the system. An example of this would be a digital circuit monitoring whether the overheated sensor has been tripped.

Inputs to a digital system are usually designated by capital letters, from the beginning of the alphabet, such as A, B, C ... or A1, A2, A2... The values taken by the digital input correspond to various discrete levels. In the case of a decimal valued input these values would range from 0 ... 9, which represents, 10 different levels. Since digital circuits require voltage or current inputs, in the case of a decimal system one would need 10 different voltage or current levels corresponding to the 10 possible decimal

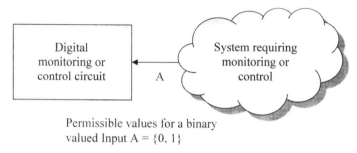

Permissible values for a binary
valued Input A = {0, 1}

Figure 1-13. Digital circuit monitoring a system

values. In practice this is rather difficult to implement. Most implementations use the much simpler binary valued system. This uses only the values of 0 or 1, and has 2 different levels. Thus only 2 different voltage or current levels are needed corresponding to the 2 different binary values. The two values are representative of any 2-state system. The values could be low-high, open-closed, up-down, light-dark, on-off, or activated-deactivated. In terms of digital systems this is often termed as logic Low-High, abbreviated 'L-H'. This is true for digital outputs as well.

Another simple implementation consists of digital circuitry which simply generates output(s), which are then applied to the system. This is shown in Figure 1-14.An example of this would be manually operated speed control of a fan for cooling an enclosure. Such a system is said to be operating in an open-loop configuration. No feedback signals are used to regulate the output.

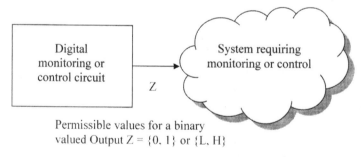

Permissible values for a binary
valued Output Z = {0, 1} or {L, H}

Figure 1-14. Digital circuit controlling a system

Outputs from a digital system are usually designated by capital letters, from the end of the alphabet, such as Z, Y, or by attaching subscripts to the output such as Z_1, Z_2, Z_3, etc.

In a more generalized implementation the operation of a system can not only be monitored by a feedback signal, it can also be controlled. This is shown in Figure 1-15.

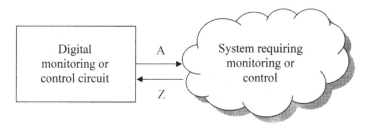

Figure 1-15. Digital circuit monitoring and controlling system

Digital circuitry is used also to process feedback information and, based on its design, generate an output which would be applied to the system. An example of this would be an air conditioning system used for cooling an enclosure. Such a system is said to be operating in a closed-loop configuration. This system has feedback signals that are used to regulate the output. This system has a Single Input, Single Output and is called a SISO system.

The monitoring and control circuitry of a system, can, in general accept multiple inputs. This includes multiple inputs from sensors and switches. The digital circuit can generate multiple outputs for switching lamps, motors, solenoids and valves. A system such as this with Multiple Inputs and Multiple Outputs is called a MIMO system. This is shown in Figure 1-16.

So far we have assumed that the system being monitored or controlled can generate signals for immediate use by the digital circuit. We also assumed that all the outputs which the digital circuit generates can be applied directly to the system. In general both input and output signals require some sort of conditioning before they can be used by the digital circuit. These changes could involve such things as simple voltage value scaling, changing between analog to digital and digital to analog formats.

A more general digital system with multiple inputs, multiple outputs, input and output conditioning circuitry as well as the monitoring and control circuit is shown in Figure 1-21.

Digital circuits are classified as being either combinational or sequential based on whether they generate their outputs based on only the present inputs, or if they make use of memory and timing functions as well.

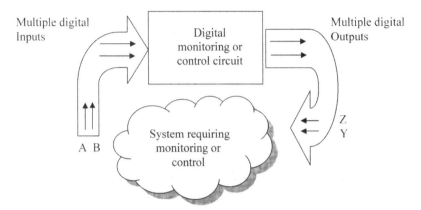

Figure 1-16. Multiple-input, Multiple-output (MIMO) digital circuit monitoring and controlling a system

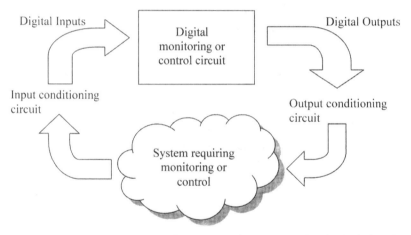

Figure 1-17. Conditioning of inputs and outputs of a digital circuit

COMBINATORIAL CIRCUITS

A digital monitoring or control circuit itself is classified as a 'combinatorial circuit' when the digital outputs depend only on the digital inputs which have been applied. This is shown in Figure 1-18. 'I' represents the values of the inputs at any given time, and 'O' represents the output values at any given time. For example, a home alarm is to be triggered when any of the sensors monitoring the fence are tripped.

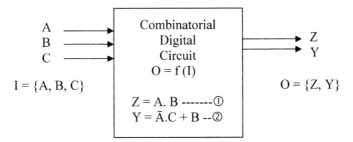

Figure 1-18. Combinatorial digital circuit

In equation ① shown in Figure 1-18, the output Z will be switched on when input A is at logic high and input B is high. This is represented using Boolean Algebra, a symbolic way of expressing and manipulating logical expressions. In this case, use is made of the AND operator '.', thus Z = AB, also written as or Z = A•B or Z = AxB.

In equation ② of Figure 1-18, use is made of the OR operator '+', along with the complement NOT operator '˜', as well as the AND. When a number of operators are present in a Boolean expression, the order of execution follows a fixed order as in arithmetic which is (sign of the number whether positive/negative, followed by exponents, then by either multiplication or division, and finally by either addition or subtraction). In Boolean algebra this order comprised of NOT, followed by AND, then by OR. Thus, the Boolean expression in ② can be paraphrased as: the output Y will be high when either 'input A is not high AND input C is high' OR input B is high.

Boolean algebra is named in honor of George Boole who first formulated it in the mid 1800s. It is a handy way of summarizing combinational logic outputs by specifying the exact conditions under which an output is switched on. This can be represented in a tabular form using a 'Truth Table' which shows all the input combinations associated with a particular output.

The Truth Table for ① of Figure 1-18 with Z = A•B, is shown in Figure 1-19.

The three logic gates NOT, AND, OR can be used to represent any combinational logic function expressed using Boolean Algebra.

The actual implementation of these Boolean expressions requires translating them into circuit symbols and interconnecting wires, then choosing the type of digital technology based on of speed, power, and space requirements.

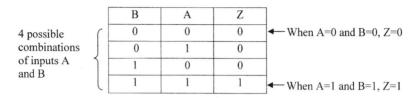

B	A	Z	
0	0	0	← When A=0 and B=0, Z=0
0	1	0	
1	0	0	
1	1	1	← When A=1 and B=1, Z=1

4 possible combinations of inputs A and B

Figure 1-19. Truth table for 'AND'

The two-input AND circuit symbol (termed as 'gate'), as designated by the IEEE (Institute of Electrical and Electronics Engineers) is shown in Figure 1-20(a). A more commonly used symbol which represents this gate is shown in Figure 1-20(b). Other symbols have been created for all logic gates.

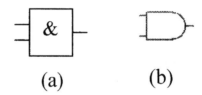

(a)   (b)

Figure 1-20. IEEE and conventional symbols for the 'AND' gate

The circuit diagram for Boolean expression ① Z = A•B is shown in Figure 1-21. It should be noted that this diagram does not show the input or output conditioning circuits, nor does it show the power and ground connections which are needed for the operation of all digital devices. It just shows the gate symbols required for implementing the circuit.

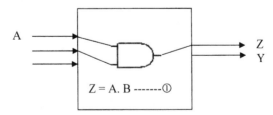

Z = A. B -------①

Figure 1-21. Digital circuit implementing the function Z = A.B

Instead of using several different types of gates for implementing different operations such as NOT, AND, OR it would be convenient to have just some general logic gates which could be used in suitable arrangements so as to perform these functions. Then one could implement any possible combinational logic function using gates of just one of this general type. In fact, such gates do exist, and for obvious reasons are termed as Universal Gates. There are two universal gates. These are the NAND and NOR gates. The NAND is obtained by connecting the output of an AND gate into a NOT. The resultant then is an AND-NOT or NAND. The circuit symbol of a NAND gate is shown in Figure 1-22.

Figure 1-22. Conventional circuit symbol for a 'NAND' gate

Combinational circuits can become extremely large even for systems having a limited number of inputs and outputs. This creates potential problems during implementation, as a large number of gates must be used and their connection points soldered. Simplification of circuits prior to implementation reduces cost of hardware, possibility of manufacturing errors, and makes troubleshooting easier. The laws of Boolean Algebra are used to simplify the final output expressions. For example, repeated ORing of a Boolean expression simply yields the same result: $Y = AB + AB + AB = AB$. It would thus be easier to implement just the expression $Y = AB$.

SEQUENTIAL CIRCUITS

A digital circuit is classified as a 'sequential circuit' if the output depends not only on the digital inputs but also on the state the system. The state of the system is updated using a state machine circuit, which is based on present inputs and the present state information. The overall output of the circuit depends on the inputs and the state of the system. This is shown in Figure 1-23. An example of this would be a display system which counts up from 0 to 9 every time a push-button is pressed and released. In order for the count to increment the system keeps track of the input (push-button) and the state (present count). In a sequential circuit, the sequencing of operations can be related only to an event with no consideration for time. In other cases the sequencing of operations can be dependent on timing.

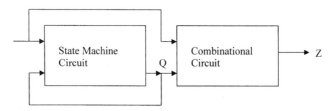

Figure 1-23. Sequential (memory/timing based) circuits

Sequential circuits can be used for:

- **Storing Digitized Values**

  The basic electronic storage or memory circuit for a bit, i.e. 0 or a 1 is a flip-flop (FF). This circuit has two stable states and is thus able to store the value of a single bit. A typical flip-flop circuit may consist of a Set and a Reset input, used for storing a 1 or 0 respectively. It also has outputs, typically two, which are complements of each other. The symbol of an un-clocked flip-flop is shown in Figure 1-24.

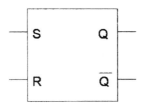

Figure 1-24. Conventional circuit symbol of a SR (set-reset) flip-flop

Many flip-flops are used along with clock signals for transferring the value appearing on the input to be stored when a clock edge (rising or falling) appears. Such inputs are termed as synchronous inputs since they are synchronized with the clock. Flip-flops may have other inputs as well for unconditionally setting and clearing regardless of the Set/Reset input and the clock edge. Such inputs are termed as asynchronous inputs. They play an important role during the startup-initialization process of a complex digital system. By clearing certain flip-flops, and setting other flip-flops as part of the startup process, the entire digital system can be placed in a specified state.

Shifting Digitized Values

When multiple flip-flops are connected together it is possible to store more than a single bit value. For example, using 2 flip-flops it is possible to store $2^2 = 4$ values. A group of flip-flops connected so as to store a digital value is called a 'register'.

Registers find use in complex digital and communication circuits. Data may arrive to a digital system over a 'serial' link. This is shown in Figure 1-25. Prior to processing the data into a serial link, it is often buffered or stored, using registers, with each arriving bit pushing the previous one into the subsequent flip-flop.

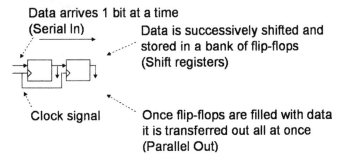

Figure 1-25. Shifting of data using flip-flops

Counting

As with registers counting circuits are built using flip-flops. The digital value which is stored in a counter can be incremented or decremented using the proper configuration of the flip-flops, inputs and a clock signal. This is shown in Figure 1-26. In some cases additional combinational circuits are used along with the sequential registers to create the counter circuit. It is used in as diverse places as digital clocks, timers in microwaves, and count-down circuits used during the launch of a spacecraft. Displaying and logging time based events uses counting circuits. The Radio Frequency Identification or RFID based timer, is used to log the time employees enter and leave, based on RF emitters embedded in the badges.

Arithmetic and Logic Operations

Data input into a system may need to be added, subtracted, multiplied, divided, by other values. Digital circuits are very well suited to this task of repeated operations for very large size numbers and over extended times. Most often the arithmetic operations are bundled along with the logic based operations in a unit called the ALU (Arithmetic and Logic Unit). Figure 1-27 shows the block diagram of an ALU.

Figure 1-26. Multi-functional timer/counter digital device

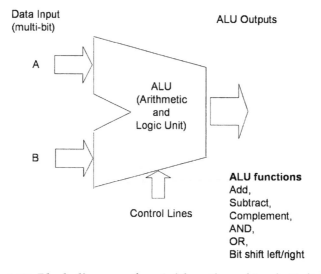

Figure 1-27. Block diagram of an Arithmetic and Logic Unit (ALU)

BLOCK DIAGRAM OF AN ARITHMETIC
AND LOGIC UNIT (ALU)

Data Conversion (Analog to Digital; Digital to Analog)

Most physical quantities are analog (continuously varying) in nature, such as temperature, pressure, flow rates. Before a digital system can act on them we need to convert these quantities into electronic signals. The device which converts the variation of a physical quantity into equivalent electrical form is designated as a transducer. In most cases the resulting electrical signal is also continuously varying. Once we have the equivalent electrical quantity it needs to be converted into digital form for processing by a digital circuit. The device which handles this conversion is an Analog to Digital Converter or ADC. With a single bit only the information regarding whether the variable is present at a designated level or absent can be recorded. Using multiple bits, it is possible to identify intermediary values between the lowest and maximum variation. The interval between which measurement samples of the signal amplitude are taken plays a part in determining the quality of the signal as well. Samples which are taken at shorter time intervals, and with multiple bits used for representing the amplitude of the signal, result in a much more accurate representation of the analog system they emulate. This information would then be provided to the digital system for processing. Once the processing is complete a digital output is produced. In cases where an analog output is needed, as in the case when the output should be proportional to the input, the digital signal must be converted into analog form. The device used for this is a DAC (Digital to Analog Converter). This analog to digital and digital to analog conversion is used extensively in sound systems, first by a microphone digitizing an audio signal using an ADC, processing it, converting the digital signal into analog form using a DAC, and finally playing it back on the speakers.

Advanced Digital Systems—Buses, Memory, Computers

Modern high-speed digital devices are capable of performing multiple functions. This includes storing/retrieving data, computing billions of operations per second, simultaneously monitoring and controlling multiple appliances, including multipurpose programmable devices for performing different tasks, and ensuring secure communications. Digital devices are thus growing progressively 'human' in their capabilities. The increase in processing speeds and the shrinking in size of hardware made it possible to fabricate such devices on a single chip.

The use of the 'bus' architecture makes it possible for multiple input and output devices to share communication lines. This reduces the number of connections required from each input/output device to the main control circuit.

Most modern day complex digital systems make use of at least three buses, one of which we have already seen is the 'data bus', which conveys data from one device to another. These devices could include the contents of primary and secondary storage locations as well. The second bus is the 'control bus' which serves to send control signals, such as enable to different devices. This is especially useful when there are a large number of devices which need to be enabled or placed in different operating modes. The third bus is the 'address bus', which is used for identifying different storage locations. A 10-bit address bus for example, can access $2^{10}=1024$, commonly referred to as 1K address locations. Once a particular address has been accessed, data can be transferred to or from it to another digital device or location. The content stored at a particular address location can be as small (1 bit) or large (128-bit), and has no relation to the actual size or format of data stored at that location.

It is common practice while sketching digital circuits which use multiple lines to denote them by using just a single wire, with a slash cutting it diagonally. The number of wires used for this purpose is indicated alongside. In Figure 1-28 the 'c#', 'a#', and 'd#' denote the number of control, address and data wires respectively. Alternatively, thicker arrows may be used for denoting multiple wires as is done in Figure 1-28. Unidirectional arrows indicate the flow of information in one direction, whereas bidirectional arrows indicate that information may flow in both directions.

In addition the task of transferring data from a particular input device to the digital circuit is often offloaded to sub-circuits, which are designated as 'input controllers'. Similarly an 'output controller' handles the task of transferring data from the main digital circuit or from memory to an output device. This is shown in Figure 1-28.

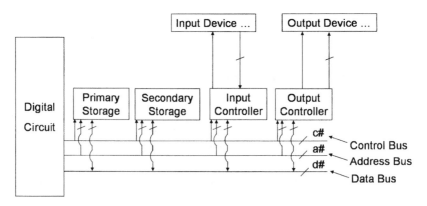

Figure 1-28. Block diagram of an advanced digital system

Data storage in advanced digital systems such as computers is commonly referred to as storage. Primary storage used for a high-speed data access, such as the cache or memory used by microprocessors. Secondary storage refers to data stored on the hard-drive and to media which may be removable. These include optical media such as Digital Versatile Disks or DVDs, Compact Disks or CDs, and magnetic media such as Thumb drives, tape drives, removable hard-drives on computer systems.

The digital circuitry used by advanced digital systems, referred to as a microprocessor, is capable of performing arithmetic and logic calculations, coordinating data transfer between devices, including storage and itself. Prior to the invention of the microprocessor this task was performed by set of ICs which functioned as the Central Processing Unit or CPU. The microprocessor is thus a CPU on a chip. Figure 1-29 shows a representative microprocessor.

In modern computers the microprocessor controls all the operations. A simplified block diagram of a microprocessor based computer system is shown in Figure 1-30.

Microprocessors with inbuilt storage and with enhanced input/output capabilities capable of interfacing with external devices are termed as microcontrollers. They are in essence a computer on a chip. With the advances in the area of digital electronics, one can expect that microcontrollers will find use in almost all technological aspects of human endeavor.

Figure 1-29. Microprocessor

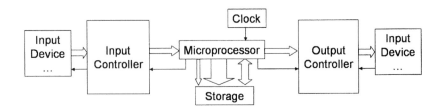

Figure 1-30. Block diagram of an earlier computer system

SUMMARY

The field of electronics studies systems whose operation can be controlled by the flow of electrons. Digital electronics controls systems where the signals can take discrete or digital values. Analog electronics controls systems where the signals can take a continuous range of values. The process of digitizing converts an analog signal to a digital one. Digital systems permit increase in storage, speed of processing, programming, and are relatively immune to interference from other signals. The world around us is largely analog and converting between analog and digital formats for processing causes delays, as well as reduces precision.

Binary valued systems use two symbols, 0 and 1, alternatively represented as low and high. Digital systems using binary valued symbols assign voltage ranges corresponding to the low and high binary values. The 2-state concept is fundamental to digital systems, which primarily use binary. A periodic waveform repeats at regular intervals called the 'time period'. The number of times a waveform repeats in a second is called the 'frequency'. Using 1 bit, 2 different values (0 and 1) can be obtained; using 2 bits 4 values (2^2); using 3 bits 8 values (2^3)—thus with each bit increase the total number of values possible doubles. A decimal system uses 10 different symbols; an octal system uses 8 different symbols and a hexadecimal system uses 16 different symbols.

Text in digital is represented in code form, such as American Standard Code for Information Interchange (ASCII) or Unicode Transformation Format (UTF). Binary Coded Decimal (BCD) uses a group of 4 binary numbers to represent an equivalent decimal digit.

Integrated Circuits (ICs) or chips incorporate resistors, capacitors, diodes and transistors on a single piece of semiconductor material. Digital devices use two types of technology Transistor Transistor Logic (TTL) for high speed, high power applications, and Complementary Metal Oxide Semiconductor (CMOS) for low power, mobile applications.

The three basic gates used in digital circuits are NOT (inverter), AND, and OR. The two universal gates, from which all other gates can be built, are the NAND and NOR. Truth tables are used to represents the output of a digital circuit in tabular form for all combinations of inputs.

A digital circuit in which the output depends only on a particular combination of inputs applied is said to be a combinational circuit. Combinational circuits are used to perform logic operations (equality), for data selection (multiplexing and demultiplexing), coding (encode and decode) data.

A digital circuit in which the output depends on a combination of the inputs applied as well as the state (memory or timing) is said to be a sequential circuit. Sequential circuits are used for storing, shifting, counting, arithmetic and logic operations, data conversion.

Advanced digital systems make use of the 'bus architecture' by sharing communication lines for data transfer between devices. High-speed storage devices are referred to as random access memory (RAM) and retain values only while powered. Storage devices which retain their values even when powered down include several types read only memory (ROM), magnetic media such as computer hard-disks, tape, flash and optical media such as compact disk (CD) digital versatile disk (DVD). Tri-state devices offer complete electrical isolation between digital system components by permitting operation in 3 states—low, high, high-impedance. The central processing unit (CPU) consists of several chips for running programs, arithmetic and logic operations, accessing input/output and storage devices. The microprocessor is a CPU on a chip. The microcomputer is a computer on a chip.

Chapter 2

Digital Logic Gates

INTRODUCTION

Digital logic systems, no matter how complex, are composed of a small group of identical building blocks. These blocks are either decision-making circuits or memory units. A large majority of the decision-making circuits are made up of logic gates or a combination of logic gates. Logic gates respond to binary input data and produce an output that is based on the status of the input. Memory circuits are used to store binary data and release it when the need arises.

Logic gates are essentially a combination of high-speed switching circuits. These gates are the electronic equivalent of a simple switch connected in series or parallel. Digital systems combine large numbers of these gates in decision-making circuits. We investigate the simple switch type of logic circuit to explain basic logic functions.

In most digital systems we do not use mechanical switch logic gates. For example, they respond very slowly to data. Electronic logic gates have been designed that can change states very quickly. In fact, these gates can change states so quickly that a human cannot detect the switching time. Typical switching times are less than a microsecond, or 10^{-6}s. In many microprocessor-based digital systems, switching times are in the nanosecond, or 10^{-9}s, range. This takes special circuits to detect a state change in logic gates operating at this speed. Logic gates respond in the same manner.

Binary Logic Functions

Any bi-stable circuit that is used to make a series of decisions based on two-state input conditions is called a binary logic circuit. Three basic circuits of this type have been developed to make simple logic decisions: the AND circuit, the OR circuit, and the NOT circuit. The logic decision made by each circuit is unique and very important in digital system operations.

Electronic circuits designed to perform specific logic functions are commonly called *gates*. This term refers to the capability of a circuit to pass or block specific digital signals. A simple if-then type of statement is often used to describe the basic operation of a logic gate. For example, if the inputs applied to an AND gate are all 1, then the output will be 1. If a 1 is applied to any input of an OR gate, then the output will be 1. If any input is applied to a NOT gate, then the output will be reversed.

The fundamental operation of a digital system is based directly on gate applications. Technicians working with digital systems must be very familiar with each basic gate function. The input-output characteristics and operation of basic logic gates serve as the basis of this discussion.

AND Gates

An AND gate is designed to have two or more inputs and one output. Essentially, if all inputs are in the 1 state simultaneously, then a 1 will appear in the output. Figure 2-1 shows a simple switch-lamp circuit of the AND gate, its symbol, and an operational table. In Figure 2-1(a), a switch turned on represents a 1 condition, whereas off represents a 0. The lamp also displays this same condition by being a 1 when it is on and 0 when turned off. Note that the switches are labeled A and B, whereas the lamp, or output, is labeled C. The operational characteristics of a gate are usually simplified by describing the input-output relationship in a table. The table in Figure 2-1(c) shows the 1 and 0 alternatives at the input and the corresponding output that will occur as a result of the input. As a rule, such a description of a gate is called a truth table. Essentially, it shows the predictable operating conditions or a logic circuit.

Each input to an AND gate has two operational states of 1 and 0. A two-input AND gate would have 2^2, or 4, possible combinations that would influence the output. A three-input gate would have 2^3, or 8, combinations, and a four-input would have 2^4, or 16, combinations. These combinations are normally placed in the truth table in binary progression order. For a two-input gate, this would be 00, 01, 10, and 11, which shows the binary count of 0, 1, 2, and 3 in order.

Functionally, the AND gate of Figure 2-1(a) produces a 1 output only when switches A and B are both 1. Mathematically, this action is described as A x B = C. This expression shows the multiplication operation. In a machine operation, this type of gate could be used to protect an opera-

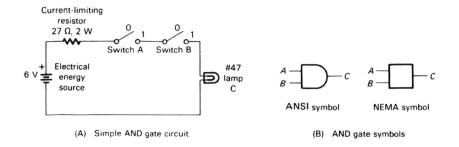

(A) Simple AND gate circuit.

(B) AND gate symbols

Switch A	Switch B	Lamp C
0	0	0
0	1	0
1	0	0
1	1	1

(C) AND gate truth table.

Figure 2-1. AND gate information: (a) simple AND gate circuit; (b) AND gate symbols; (c) AND gate truth table.

tor from some type of physical danger. For example, it will not permit a machine to be actuated until the operator presses one button with the left hand and a second button with the right hand at the same time. This removes the hands from a dangerous operating condition.

The symbol representations of an AND gate shown in Figure 2-1(b) are very common. The symbol on the left side has been adopted by the American National Standard Institute (ANSI) and the Institute of Electrical and Electronic Engineers (IEEE). The symbol on the right side is used by the National Electrical Manufacturer's Association (NEMA). Both symbols are in common use today.

OR Gates

An OR gate is designed to have two or more inputs and a single output. Like the AND gate, each input to the OR gate has two possible states: 1 and 0. The output of this gate will produce a 1 when either or both inputs are 1. Figure 2-2 shows a simple lamp-switch analogy of the OR gate, its symbol, and a truth table.

Functionally, an OR gate will produce a 1 output when both switches are 1 or when either switch A or B is a 1. Mathematically, this action is described as $A + B = C$. This expression shows OR addition. Applications of this gate are used to make logic decisions as to whether or not a 1 appears at either input. The interior light system of an automobile is controlled by an OR type of circuit. Individual door switches and the dash panel switch all control the lighting system from a different location. Essentially, when any one of the inputs is on, it will cause the interior lights to be on.

NOT Gates

    A NOT gate has a single input and a single output, which makes it unique compared with the AND and OR gates. The output of a NOT gate is designed so that it will be opposite to that of the input state. Figure 2-3 shows a simple switch-controlled NOT gate, *its* symbol, and truth table. Note that when the single-pole, single-throw (SPST) switch is on, or in the 1 state, it shorts out the lamp. Likewise, placing the switch in the off condition causes the lamp to be on, or in the 1 state. NOT gates are also called inverters. Mathematically, the operation of a NOT gate is expressed as A', A^*, or $\bar{A}$. The $\bar{A}$ or (A-bar) symbol shows the inversion function. The significance of a NOT gate should be rather apparent after the following discussion of gates.

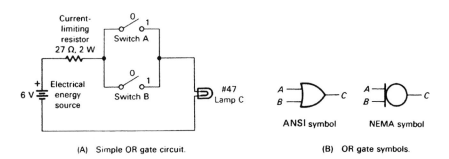

(A)   Simple OR gate circuit.

(B)   OR gate symbols.

Switch A	Switch B	Lamp C
0	0	0
0	1	1
1	0	1
1	1	1

(C)   OR gate truth table.

Figure 2-2. OR gate information: (a) simple OR gate circuit; (b) OR gate symbols; (c) OR gate truth table.

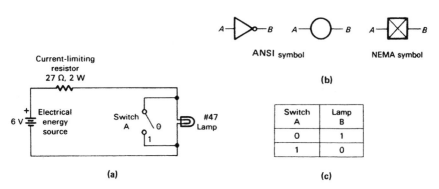

Figure 2-3. NOT gate information: (a) simple NOT gate circuit; (b) NOT gate symbols; (c) NOT gate truth table.

Combination Logic Gates

When a NOT gate is combined with an AND gate or an OR gate, it is called a *combination logic function*. A NOT-AND gate is normally called a NAND gate. This gate is an inverted AND gate, or simply NOT an AND gate. Figure 2-4 shows a simple switch-lamp circuit analogy of this gate, along with its symbol and truth table.

The NAND gate is an inversion of the AND gate. When switches A and B are both on, or in the 1 state, the lamp C is off. When either or both switches are off, the output, or lamp C, is in the on, or 1 state. Mathematically, the operation of a NAND gate is expressed as $A \times B = C^*$. The C^* denotes the inversion, or negative function, of the gate.

A combination of NOT-OR, or NOR, gate produces an inversion of the OR function. Figure 2-5 shows a simple switch-lamp circuit analogy of this gate, along with its symbol and truth table. When either switch A, B, or A and B are off, or 0, the output is a 1, or high. When either switch A or B is 1, the output is 0. Mathematically, the operation of a NOR gate is expressed as $A + B = C^*$. A 1 will appear in its output only when A is 0 and B is 0. This represents a unique logic function.

Logic Gate Circuits

The switch-lamp analogy of gate circuits was used primarily to show basic logic operating characteristics in simplified form. In actual circuit applications, logic gates are very seldom built with switches and indicating lamps. Transistors, diodes, and a variety of other components are connected together to achieve different gate functions. The switching action of a gate can then be achieved very effectively by applying either forward

Figure 2-4. NAND gate information: (a) NAND gate circuit; (b) NAND gate symbols; (c) NAND gate truth table.

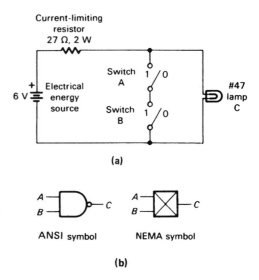

or reverse biasing to these solid-state components.

The component structure of a gate is very important when selecting a gate for a specific circuit application. In this regard, the components of the circuit have a great deal to do with the response of the gate. Some components have a storage characteristic that will delay the switching time of a circuit. Specific components also have a power dissipation rating that must be taken into account. Diodes, bipolar transistors, MOSFETs, and component coupling are some of the things that must be taken into account when selecting components for a logic gate.

Switch A	Switch B	Lamp C
0	0	1
0	1	1
1	0	1
1	1	0

(c)

Diode Logic AND Gate

A simple but very useful electronic gate utilizes the switching properties of diodes. A diode is an electronic device that conducts current when forward biased. When a diode is conductive, the voltage across it drops to a very low value. For a silicon diode the forward voltage is 0.7 V. When the same diode is reverse biased, it does not conduct. The voltage across it equals or approximates the source voltage. The diode therefore acts like a switch. The switch is closed when it is forward biased and open when reverse biased.

A two-input diode resistor AND gate is shown in Figure 2-6. In this circuit logic signals are applied to the input through switches A and B. In

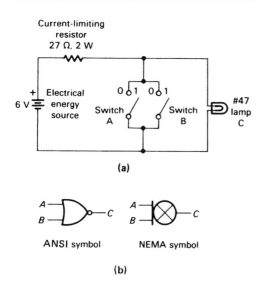

Figure 2-5. NOR gate information: (a) NOR gate circuit; (b) NOR gate symbols; (c) NOR gate truth table.

(a)

(b)

Switch A	Switch B	Lamp C
0	0	1
0	1	0
1	0	0
1	1	0

(c)

practice, input logic signals are not necessarily produced by mechanical switches. An actual circuit would derive its input from other logic circuits. The output of the gate appears across resistor R_L. Output voltage is an indication of the high, or 1, state, whereas no voltage represents the low, or 0, state.

Operation of the two-diode AND gate is based on the position of switches A and B. If a switch is in the high position, it will connect the positive side of the source to the cathode of a respective diode. In the low position the negative side of the source will be connected to the cathode of a respective diode. The switch simply causes a diode to be forward biased when low, or negative, and reverse biased when high, or positive. Four possible input combinations can be achieved by this circuit.

Assume now that both switches of the diode logic gate are in the low position, as indicated in Figure 2-6. This condition forward-biases both diodes. Current will flow through D_1, D_2, and resistor R_L. A voltmeter connected to point C or the output will show the voltage drop across each diode. In this case, the voltage will be 0.7 V. This represents a low, or 0, state of the logic gate. With switch A low and switch B low, the output will be low. The truth table of Figure 2-6(b) shows this as step 1.

Referring again to Fig 2-6, let us see how the logic gate responds to other switch positions. Step 2 of the truth table shows switch A low and switch B high. This condition causes diode D_1 to be forward biased and D_2

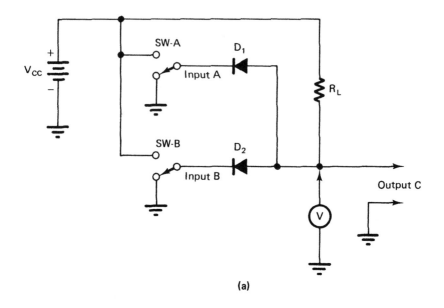

(a)

Operational steps	Inputs		Output
	A	B	C
1	Low	Low	Low
2	Low	High	Low
3	High	Low	Low
4	High	High	High

(b)

Figure 2-6. Two-input diode AND gate.

to be reverse biased. Current conduction through D_1, the forward-biased diode, will cause the voltmeter to indicate a low, or 0, state. The reverse-biased diode D_2 will not alter the current flow through R_L. A 0, or low, at the input of D_1 and a high, or 1, at D_2 will cause a zero to appear at the output.

Step 3 of the truth table of Figure 2-6 shows how the diodes will respond when input A is high and input B is low. In this case, diode D_1 is reverse biased and diode D_2 is forward biased. Current conduction through D_2 will cause the voltmeter to indicate a low, or 0 state. The reverse-biased diode D_1 will not alter the current flow through R. A high state, or 1, at the input of D_1 and a low, or 0, at the input of D_2 will cause a zero to appear at the output.

Step 4 of the truth table of Figure 2-6 shows how the diodes will respond when input A is high and input B is high. In this case, both diodes are reverse biased. No current will flow through a reverse-biased diode. As a result, the voltage across each diode will equal the value of the source. If 5 V are used as the source, the output will be 5 V. The voltmeter, connected to point C, will indicate this value. This means that a high state, or 1, appearing at both inputs A and B will cause a 1, or high, to appear at the output.

Diode Logic OR Gate

The OR function can be achieved with diodes connected in the circuit configuration of Figure 2-7. Compared with the AND gate of Figure 2-6, the two diodes are reversed and the load resistor is attached to the diode connection point and the ground. A truth table of the logic circuit is included with the circuit diagram.

In step 1 of the truth table, the input to each diode is 0. This reverse biases the two diodes. With a reverse-biased diode, no conduction takes place. As a result, the voltage measured across the load resistor is zero or low. This means that if A or B is 0, then the output will be 0.

Step 2 of the diode OR gate has input A low, or 0, input B high, or 1, and the output high, or 1. When input B is high, it forward-biases diode D_2. D_1 is reverse biased when it is low. This condition causes current to flow through the load resistor because of the conducting diode. As a result, there is a voltage drop across R_L. The value of this voltage is the source voltage less 0.7 V across a conducting diode. For a Vcc of 5 V, the output would be 5 V — 0.7, or 4.3 V.

Step 3 is primarily the same as step 1 for an OR gate. In this case, the polarity of the voltage applied to the two diodes is simply reversed. D_1 is conductive or forward biased, and D_2 is reverse biased. This causes current to flow through R_L. A voltage drop appears across R_L, indicating a high, or 1, output. The voltage value is the source voltage less 0.7 V across a conducting diode.

Step 4 shows the same response as steps 1 and 2. In this case, both diodes are forward biased at the same time. This again causes current flow through the load resistor and voltage to appear across it. The voltage value is primarily the same as the previous two steps. Step 3 has input A high, input B high, and the output high.

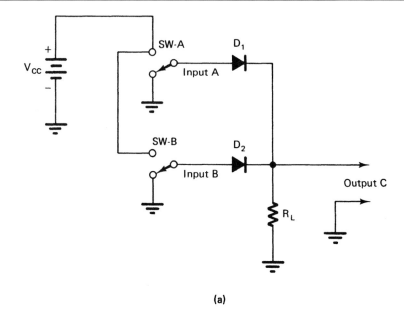

(a)

Operational steps	Inputs		Output
	A	B	C
1	0	0	0
2	0	1	1
3	1	0	1
4	1	1	1

(b)

Figure 2-7. Two-input diode OR gate.

Transistor-Transistor Logic

Transistor-transistor logic (TTL) is a refinement of the DTL family. In TTL, a single multiemitter transistor replaces the input diodes and series diode of the DTL type of device. Each emitter-base junction serves as an input, and the base-collector junction serves as a series diode. Output of the first transistor is coupled directly to second transistor. The overall operation of a TTL gate is very similar to that of the DTL family. Figure 2-8 shows the circuitry of a basic TTL NAND gate.

Notice the structure of the multiemitter transistor T_1. This is the distinguishing feature of a TTL gate. In this case, the transistor has three emitters. Each emitter serves as an independent input. The combined input signal drives the base-collector junction. The output of T_1 drives the base of transistor T_2. This transistor inverts the signal applied to its input.

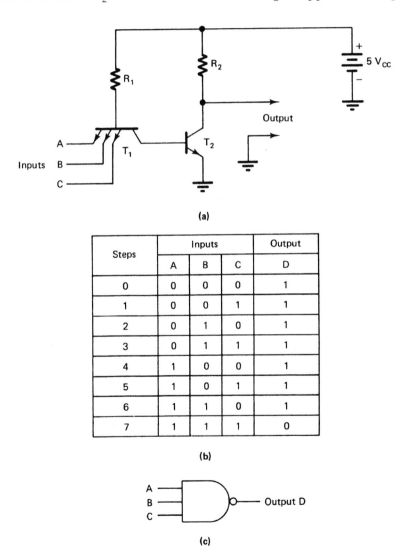

(a)

Steps	Inputs			Output
	A	B	C	D
0	0	0	0	1
1	0	0	1	1
2	0	1	0	1
3	0	1	1	1
4	1	0	0	1
5	1	0	1	1
6	1	1	0	1
7	1	1	1	0

(b)

(c)

Figure 2-8. A three-input TTL AND gate: (a) circuit; (b) truth table; (c) NAND gate symbol.

These transistors are directly coupled with the output of the gate appearing at the collector of T_2.

Suppose now that all inputs of the TTL gate of Figure 2-8 are at a high (or logic 1) state. This voltage would approximate the value of Vcc. A positive voltage at the emitter of an NPN transistor causes it to be reversed biased. All three inputs are reverse biased at the same time. Biasing of T_1 is such that the base-collector junction is forward biased by resistor R_1. With the emitters all reverse biased, the base-collector is forward biased. This condition causes current flow through R_1 and the base of T_2. This drives T_2 into saturation. The output of T_2 is nearly 0 V or at the same level as ground. The truth table of Figure 2-8(b) shows this condition as step 7. This is representative of the response of a NAND gate.

If any one or any combination of the three inputs is made 0 (or low), it will change the output of the NAND gate. A low-level input applied to any emitter of T_1 will forward bias it into conduction. This causes an increase in current through R_1. A larger current flow causes increased voltage drop across the resistor and reduces the voltage at the base of T_2. The change in base voltage is great enough to reverse bias T_2. The collector voltage of a reverse-biased transistor rises to the value of the source voltage. In this case, the output approximates the value of V_{CC}. A low (or 0) at any one of the inputs will cause the output to be high, or in the 1 state. Steps 0 through 6 show this operation in the truth table.

Emitter-Coupled Logic

The shortest delay time available today is achieved by using emitter-coupled logic, or ECL gates. The circuitry of this logic family is designed so that the collector-emitter voltage of the transistors is always above 0.3 V. This prevents the transistors from going into heavy saturation. The storage charge of a transistor is reduced and gate delay is shortened. The fastest available ECL gates have a delay of 1 ns for a power dissipation of 60 mW per gate. In other versions of the ECL series, delay times are 2 ns for a power dissipation of 25 mW per gate. This logic family has a definite trade-off between delay time and power dissipation.

Figure 2-9 shows the circuitry of an ECL NOR gate. The emitters of logic transistors T_1, T_2, and T_3 are coupled to the emitter of reference transistor T_4. The current flow through the common-emitter resistor R_E is maintained at a constant value. The base of transistor T_4 is connected to a reference voltage source. If the three inputs are all near ground or 0 V, it causes T_1, T_2, and T_3 to be cut off. This stops the current from passing

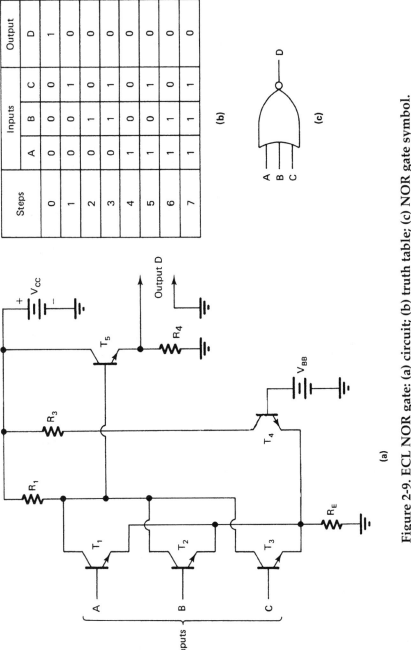

Steps	Inputs			Output
	A	B	C	D
0	0	0	0	1
1	0	0	1	0
2	0	1	1	0
3	0	1	1	0
4	1	0	0	0
5	1	0	1	0
6	1	1	0	0
7	1	1	1	0

(b)

(c)

(a)

Figure 2-9. ECL NOR gate: (a) circuit; (b) truth table; (c) NOR gate symbol.

through the collector resistor R_1. The voltage across R_1 rises to the value of V_{CC}. A high positive voltage at the base of T_5 drives it into full conduction. Current flow through R_4 causes the output to be positive and produce a logic 1. This represents step 0 of the truth table.

If any one of the inputs is made positive and above the value of V_{BB}, it will cause a specific transistor to be conductive. The collector voltage across R_1 will drop in value. This in turn causes transistor T_5 to turn off and produce a 0 (or low) output. Since the resistance of R_3 maintains the current at a constant value, an increase in logic transistor current lowers the current flow through reference transistor T_4. The switching threshold voltage of a logic transistor is equal to the reference voltage source. The use of emitter coupling in the logic transistor stage and the output stage prevents these transistors from going into saturation. This essentially causes the switching speed to be very fast. The logic levels of this gate are not as well defined as those of other logic families. The output impedance is quite low, which permits a number of external gates to be connected to the output.

CMOS Logic

CMOS logic was designed for low-power dissipation circuit applications. This logic family uses N-channel and P-channel MOSFETs connected in a series-circuit configuration. The polarity and value of the applied gate voltage causes respective transistors to be conductive. A positive voltage applied to the gate of a P-channel FET drives the current carriers away from the channel causing non-conduction. A negative voltage of sufficient value will pull current carriers into the channel, thus causing it to be conductive. The N-channel transistor responds in the same way except that it is conductive with positive voltage and off with negative voltage. Without input voltage applied to the two transistors, very little current is consumed from the source.

Figure 2-10 shows a circuit diagram, truth table, and the logic symbol of a CMOS logic inverter. Input is applied to the gate of each transistor and output is developed across the N-channel device. When the input is at approximately 0 V, Q_1 is turned off and Q_2 is made conductive. This action connects Vcc to the output, which is in its 1 (or high) logic state. When the input is at + 5 V, Q_2 is turned off and Q_1 is made conductive. This causes the output to be low (or 0). The truth table shows the output as an inversion of the input.

The transition between on and off states of the CMOS logic family

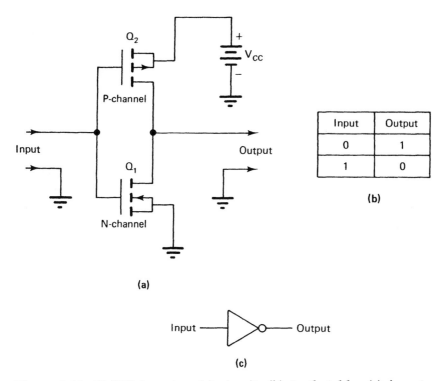

Input	Output
0	1
1	0

(b)

(a)

(c)

Figure 2-10. CMOS inverter: (a) circuit; (b) truth table; (c) inverter symbol.

is somewhat more gradual than that achieved by TTL or DTL logic families. A more pronounced state change can be achieved by connecting the output on an inverter to the input of a duplicate inverter. The output of the second inverter is then connected to a third inverter. This circuit modification causes the composite inverter to have a sharp transition from on to off that is comparable to that of the TTL family. An inverter of this type is considered to be buffered. The resulting output of a buffered inverter is the same as that expressed by the truth table of an un-buffered inverter. Buffering generally increases the delay time of the switching operation.

When a CMOS logic gate is in one of its operational states (high or low), the only current that flows is leakage current. Typical current values are approximately 10 μA. With a supply voltage of 5 V, a gate would dissipate 5 V x 10 μA or 50 μW. This low-power-dissipation characteristic is an important selection consideration for digital systems that are energized by a small source. CMOS can also be used with voltage source values up to 15 V. Portable logic circuits energized by 12-V, low-power digital watch circuits operating from a 1.25-V battery; and solar powered pocket calculators are common applications of this logic family.

SUMMARY

Three basic gates have been developed to make logic decisions. These are AND, OR, and NOT. The functional operation of a logic gate is based on a simple if-then type of statement. If the inputs of an AND gate are all 1, then the output will be 1. If a 1 is applied to any input of an OR gate, then the output will be a 1. If any input is applied to a NOT gate, then the output will be reversed.

An AND gate is designed to have two or more inputs and one output. If all inputs are 1, then the output will be 1. If any input is 0, then the output will be 0. Mathematically, the action of an AND gate is expressed as $A \times B = C$. This is the multiplication operation.

An OR gate is designed to have two or more inputs and a single output. Functionally, an OR gate will produce a 1 output when a 1 appears at any input. Mathematically, this action is expressed as $A + B = C$. This is called OR addition.

A NOT gate has a single input and a single output. The NOT function is achieved by an inverter. A 1 input causes the output to be 0 and a 0 input causes the output to be 1. Mathematically, the operation of a NOT gate is expressed as $A = \bar{A}$.

When two of the basic logic gates are connected together, they form a combination logic gate. The two most common combination logic gates are NOT-AND and NOT-OR. These are generally called NAND and NOR functions.

A NAND gate is an inversion of the AND function. When a 1 appears at all inputs the output will be 0. When a 0 appears at any input the output will be 1. Mathematically, the operation of a NAND gate is expressed as $A \times B = \bar{C}$.

A NOR gate produces an inversion of the OR function. When the inputs are all 0, the output is 1. When any input is 1, the corresponding output is 0. Mathematically, the operation of a NOR gate is expressed as $A + B = \bar{C}$.

TTL employs a multi-emitter transistor to serve as the diode input logic unit. The collector-base junction of this transistor serves as the series diode. The output of the input transistor is directly coupled to the input of a second transistor.

ECL, or emitter-coupled logic, has the shortest delay time available today. The collector-emitter voltage of this IC family is maintained at 0.3 V to prevent the transistors from going into full saturation. This reduces the storage charge and gate delay is reduced. Power dissipation per gate is generally quite high.

Complementary metal-oxide silicon, or CMOS, logic was developed for low-power-dissipation circuits. This logic family uses N-channel and P-channel MOSFETs connected in a series circuit across the source. The low-power-dissipation characteristic is an important selection consideration for choosing the CMOS logic family for a circuit application.

Chapter 3

Boolean Algebra and Logic Gates

INTRODUCTION

Digital electronics operates on a mathematical base as do other forms of electronics. The operating principle of a digital electronic system is quite unique when compared with other electronic functions. Digital electronics, for example, is concerned with two-state information represented by bits. A bit may assume either one of two values: 0 or 1. The operational base of this system is considered to be binary, or base 2. Boolean algebra is the mathematical basis of digital logic because it responds to base 2 information.

Boolean algebra is an important tool when used in the digital electronics field. It is used to express logic functions in the form of an equation, to analyze data, to simplify logic expressions, to design logic circuits and to troubleshoot a logic circuit. A person working with digital logic circuits continually uses Boolean algebra.

USING BOOLEAN ALGEBRA

In 1854 George Boole, an Englishman, published a paper titled "An Investigation of the Laws of Thought." In this publication, Boole symbolized the form of algebra that deals with two-state logic. In 1938 Claude Shannon, an American, wrote a paper titled "A Symbolic Analysis of Relay and Switching Circuits." This paper applied Boole's ideas to relay switching used in telephone circuitry. Today, this is called *Boolean algebra*. It is based on standard algebraic principles that deal with the mathematical theory of two-state logic. The notations of Boolean algebra and theorems are fundamental in the study of digital electronics.

Boolean algebra is a special form of algebra that was designed to show the relationships of two-state variable logic operations.

This form of algebra is ideally suited for the analysis and design of binary logic systems. Through the use of Boolean algebra, it is possible to write mathematical expressions that describe specific logic functions. Boolean expressions seem to be much more meaningful than complex word statements or elaborate truth tables. The laws that apply to Boolean algebra may also be used to simplify complex expressions. Through this type of operation, it may be possible to reduce the number of logic gates needed to achieve a specific function before it is built.

In Boolean algebra the variables of an equation are commonly assigned letters of the alphabet. Each variable then exists in states of 1 or 0 according to its condition. The 1, or true, state is normally represented by a single letter such as A, B, or C. The opposite state or condition is then described as 0, or false, and is represented by A or A'. This is described as NOT A, A negated, or A complemented.

Boolean algebra is somewhat different from conventional algebra with respect to its mathematical operations. Boolean operators are concerned only with multiplication, addition, and negation. These are expressed as follows.

Multiplication: A AND B, AB, $A \times B$
Addition: A OR B, $A + B$
Negation or complementation: NOT A, $\bar{A}$, A^*, A'
Equivalency: A equals B, $A = B$

Boolean operators can be achieved with logic gates. An AND gate is used to accomplish multiplication. OR addition is achieved with an inclusive OR gate. Negation, or complementation, is achieved with an inverter, or NOT, gate.

Boolean operators play an important role in the function of a digital system. We have looked at operator functions with respect to how they are accomplished with switches, diodes, transistors, and ICs. The internal structure of the gate is not nearly as important as the function that it accomplishes. We cannot change a single transistor in an IC if it is not functioning properly. We can, however, change the input of an operator that will alter its output. It is important therefore, to recognize the function of an operator and to see what it can be used to accomplish.

Logic gates and Boolean operators up to this point have been viewed as symbol representations. It is important to be able to recognize each symbol and have some idea as to what it will accomplish. We must be able to recognize word statements that describe a logic function and be able to translate the statement into gate operations. We must be able to recognize the symbol and the statement and be able to translate these into mathematical operators. This is needed to simplify logic statements and to determine if a logic circuit is functioning properly.

Translating Digital Logic Circuits to Boolean Expressions

One important application of Boolean algebra is to express logic functions mathematically. A Boolean expression is widely used to show how an equation responds with respect to its input and output. In this regard, the inputs are assigned certain letter designations and the output is expressed as a different letter. The letter designations do not follow any pattern or sequence and can be in any order. The inputs of an AND gate could be A and B and the output could be X. The inputs and the outputs can assume either a 0 or a 1 state. The output is an expression of the logic operator achieved by the gate with respect to its input. Figure 3-1 shows how logic symbols are used to accomplish Boolean operators. Note that the operator is expressed by a logic symbol and a mathematical statement.

Being able to recognize the Boolean function of a logic gate and to express it mathematically is an important working tool. This procedure can be mastered with a little practice. Refer to Figure 3-2. First locate the logic gate being defined. Then identify the input and output letter designations.

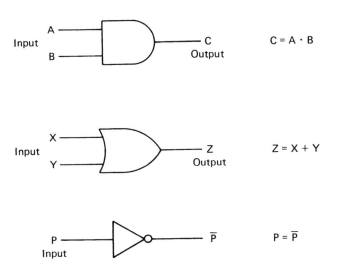

Figure 3-1. Logic symbols used to accomplish Boolean operators

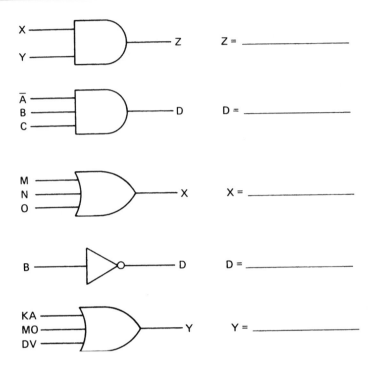

$X =$ _____

$D =$ _____

$X =$ _____

$D =$ _____

$Y =$ _____

Figure 3-2. Mathematical expressions of a logic gate

Then identify the Boolean operator of the logic gate. This is multiplication for an AND gate, addition for an OR gate, and negation for a NOT gate. Finally, combine the input, output, and operator in a statement, as shown on the right of the symbol.

When one logic gate is used to accomplish a function, it is considered to be a single-level operator. Two gates connected in a combination are considered a two-level function. The output of a two-level function is a combination of two single-level operator functions. These logic circuits are generally used to accomplish more sophisticated decision making functions. Many combination logic circuits are used in digital equipment. The functions achieved by two-, three-, and four-level operators occur quite frequently.

Two basic forms of combination logic that respond as two-level operators are called the *product of sums* and the *sum of products*. The term product refers to the multiplication operator, which is achieved by an AND gate. Sum refers to the addition operator and is achieved with an OR gate.

A product of sums circuit has the output of two or more OR gates connected to the input of an AND gate. The sum of products circuit has the output of two or more AND gates applied to the input of an OR gate. The output of a product of sums combination is $A + B \times C + D$. The output of a sum of products circuit is $AB + CD$. These logic functions are frequently used Boolean expressions.

To prepare a Boolean statement describing the function of a multiple-level logic circuit requires several procedural steps. Refer to the two-level logic circuit of Figure 3-3. Step 1 involves identification of the inputs. This generally starts on the upper left side of the symbol diagram. For this part of the diagram the inputs are A and B. Step 2 identifies the operator of the gate. The AND function or multiplication operator is achieved by this gate. Step 3 deals with the formation of a partial statement showing the input-operator function of the gate. This is AB. Step 4 repeats the procedure for each of the other first-level gates. In this circuit only one other first-level gate is used. It produces CD. Step 5 identifies the next level of gates. This is located on the right side of the first logic level. The partial statement, or output, of a first-level gate serves as an input for the second-level gate. Step 6 labels the partial input statement of the second-level gate and identifies the operator achieved by it. The inputs are AB and CD. The operator is the OR function. Step 7 prepares an output statement that shows the combined operators for each gate of level one and two. For this circuit $X = AB + CD$ is the resulting output statement. This describes the function of a combination circuit that achieves the sum of the products.

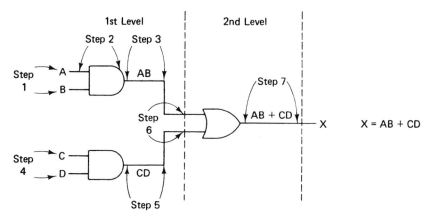

Figure 3-3. Boolean statement identification

To develop a Boolean statement for a given logic circuit takes some practice. Refer to the logic circuits in Figure 3-4. Develop a Boolean statement for each circuit. The step procedure just outlined should be used on these circuits. For a three-level logic circuit a partial statement for level one is formed first. Then a partial statement for the second level is formed. The third level combines the previous level two statements into a level three output. The resulting output statement for each circuit is included with the answers for the following self examination test.

The given Boolean Expression is:

A + B + C = X

Step 1: Identify the inputs and outputs of the statement

A + B + C = X

Inputs   Output

Step 2: Identify the primary logic operator

A + B + C = X

Primary logic operator is addition

Building Logic Circuits from Boolean Expressions

Boolean expressions are frequently used to guide a person in the building of logic circuits. Assume now that you are given the task of constructing a logic circuit from a given Boolean expression. Suppose that the given expression is $A + B + C = X$. The statement reads inputs *A* OR *B* OR *C* equals output *X*. Looking at the expression shows that inputs A, B, and C are each ORed together to equal output X. To develop a logic circuit that will accomplish this operation, we must simply select a gate that will add the inputs together. An OR gate will accomplish

Step 3: Select a logic gate that will achieve the operator function. This gate must be capable of accomodating all the inputs

Step 4: Assign inputs and outputs to the logic gate. Label the gate inputs and outputs

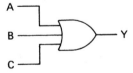

Figure 3-4. Conversion of a Boolean expression to a logic gate circuit

this Boolean operation. The OR gate must accept three inputs and develop one output. This could be accomplished by using a single-level three-input logic gate. Figure 3-4 shows the steps for solving this problem.

Now suppose that you are to build a logic circuit for Boolean expression that is somewhat complex. Figure 3-5 shows the steps for solving this problem. The statement to be defined reads $AB + A\bar{B} + C = Y$ Step 1 identifies the inputs and output of the statement. Step 2 identifies the primary logic operator of the statement. In this case the primary operator is addition. Step 3 selects a logic gate that will achieve the primary operator function. An OR gate will accomplish this function. Draw a logic symbol that will achieve this operation. Step 4 assigns input and output labels to the logic gate. The inputs are AB, $A\bar{B}$, *and* C. The output is labeled Y Step 5 identifies the logic operator achieved by each input. In this expression one input does not involve an operator. The other two inputs are each ANDed together. Step 6 selects a logic gate that will accomplish the multiplication operation. An AND gate is used for this operator function. Step 7 assigns labels to the secondary input gates. Step 8 supplies additional operators if they are needed to complete the statement. In this case, one of the secondary inputs is inverted. A NOT gate is attached to its input. Step 9 connects each of the secondary inputs to a common connection point. The Boolean expression has been converted into a usable logic circuit.

TRUTH TABLES AND BOOLEAN ALGEBRA

Boolean expressions are frequently used to describe how a logic circuit operates. We have used these expressions to construct circuits and to define operations. A logic circuit has several alternatives that must be taken into account when evaluating its operation. A graphic display of the operational steps and all its possible alternatives is sometimes needed to understand fully the operation of an expression. A display that shows these operations is called a truth table. A truth table is defined as a tabular listing of all the possible logic-level combinations produced by the input and output of a digital circuit. This means that a truth table is a specifications sheet that describes the exact behavior of a logic circuit. The circuit can have several inputs and one or more outputs.

Truth tables are constructed to show the relationship of the output of a logic circuit with respect to its inputs. A truth table lists every possible combination of the input and its corresponding output. If a logic cir-

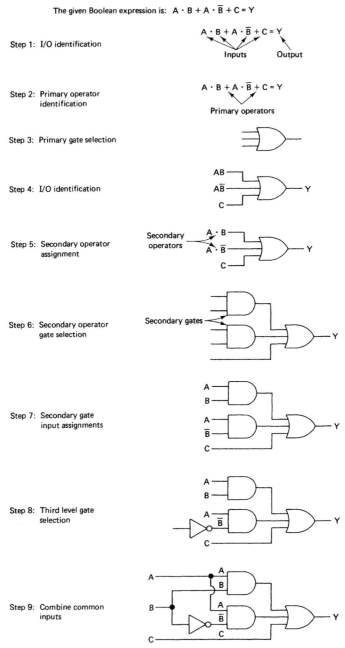

Figure 3-5. Conversion of a Boolean expression to a logic gate

cuit has 1 input, then there will be two possible combinations listed in the table. A circuit with two inputs will have four possible combinations. The number of combinations is 2 to the power of the number (N) of inputs, or 2^N. For a one-input gate, this is 2^1, or 2. For a two-input gate, this is 2^2, or 4 combinations. For three inputs, it is 2^3, or 8. Truth tables can have combinations of 2, 4, 8, 16, 32, 64, 128, and larger.

The input combinations of a truth table should be listed in logical order to avoid using the same entry twice or possibly omitting an entry. The inputs are generally listed in columns on the left side of the table. The most significant variable (input) should be listed in the left most column. The inputs are generally arranged alphabetically from left to right, such as *ABCD* or *XYZ*. However, the variables can be listed in reverse order (*DCBA* or *ZYX*) and still be the same.

In a truth table the 1 and 0 alternatives of each input are generally listed in binary progression. Binary progression refers to values that increase in size starting from 0. The first horizontal row would be 0000. The next row would be 0001. The succeeding rows would increase in binary order with values of 0010, 0011, 0100, 0101, 0110, 0111, 1000, 1001, 1010, 1011, 1100, 1101, 1110, and 1111 for a four-input truth table. A three-input truth table would have a listing of 000 to 111. A two-input table would be 00, 01, 10, and 11. Figure 3-6 shows four different truth table listings. Review the input alternatives of these tables.

Truth Table Construction

Truth tables are used to show the actual operational states of a logic circuit. They can be used to show the operation of a single logic gate or a combination of several gates connected in a complex Boolean statement. We have used truth tables to define the function of single-level logic gates. Combination logic circuits generally require some modification of the truth table to accommodate all its operations. This procedure is called *truth table construction*. In a sense, it shows the mathematical operations achieved by a Boolean expression in tabular form. There is a wide range of different combinations that can be accomplished. These can be achieved with addition, multiplication, and inversion operators.

The construction of a truth table for a single-level Boolean statement is rather easy. We simply look at the 1-0 input alternatives and perform the mathematical function of the designated Boolean operator. The output is the mathematical answer for a particular input alternative. It is indicated

Four-input truth table

Inputs				Output
D	C	B	A	X
0	0	0	0	0
0	0	0	1	0
0	0	1	0	0
0	0	1	1	0
0	1	0	0	0
0	1	0	1	0
0	1	1	0	0
0	1	1	1	0
1	0	0	0	0
1	0	0	1	0
1	0	1	0	0
1	0	1	1	0
1	1	0	0	0
1	1	0	1	0
1	1	1	0	0
1	1	1	1	1

Three-input truth table

Inputs			Output
C	B	A	X
0	0	0	0
0	0	1	1
0	1	0	1
0	1	1	1
1	0	0	1
1	0	1	1
1	1	0	1
1	1	1	1

Two-input truth table

Inputs		Output
B	A	X
0	0	0
0	1	0
1	0	0
1	1	1

Inverter truth table

Input A	Output $\overline{A}$
0	1
1	0

Figure 3-6. Truth table construction

in the same horizontal row as the input alternatives and listed in the output column.

Figure 3-7 shows some truth tables for single level logic statements. Table 1 shows the truth table for an AND operator, which involves multiplication. The output is the product of inputs A and B. The first alternative is 0 x 0 = 0. The second alternative is 0 x 1 = 0. The third alternative is 1 x 0

= 0. Alternative four is 1 x 1 = 1. Multiplication is performed horizontally on the input alternatives. The resulting answer is listed in the output column.

Table 2 of Figure 3-7 shows the truth table for a single-level OR operator, which involves addition. The output is the sum of inputs A and B. The alternatives are 0 + 0 = 0, 0 + 1 = 1, 1 + 0 = 1, and 1 + 1 = 1. Addition is performed horizontally on the input alternatives. The resulting answer is listed in the output column.

Table 3 of Figure 3-7 shows the truth table for a single-level NOT operator, which involves negation. The output is an inversion of the input. The alternatives are 0 = 1 and 1 = 0. Negation is shown horizontally by inverting the input alternative. The resulting answer is listed in the output column.

Construction of a truth table for a Boolean expression with two or more operators requires some modification of the basic truth table. Essentially, the mathematical operators of the expression are performed on inputs and listed in a column. The column is a tabulation of the different input alternatives. The number of operators in an expression determines the functional steps of the truth table. The construction procedure

Table 1

$A \cdot B = C$

Inputs		Output
B	A	C
0	0	0
0	1	0
1	0	0
1	1	1

Table 2

$A + B = X$

Inputs		Output
B	A	X
0	0	0
0	1	1
1	0	1
1	1	1

$A = \overline{A}$

Input	Output
A	$\overline{A}$
0	1
1	0

Figure 3-7. Truth tables

of a truth table is listed in the following steps.

1. First look at the Boolean expression and determine the number of inputs that are used.

2. Make a column for each input in the table. List the alternatives of the inputs in binary progression order.

3. Start with the first operator function on the left side of the Boolean expression.

4. Perform the mathematical operation for this step by using the appropriate input alternatives.

5. Prepare a column that lists the output for the operator. The operator is performed horizontally with its answers listed in a vertical column. Label the column showing the performed operator function.

6. Solve the next operational step to the right. This may involve some intermediate step before the general operator can be performed. Negation is a common step that must be performed on one or more of the inputs. Then perform the original operational step. It should include all the alternatives for this operation. The operation is performed horizontally and the resulting output is listed in a vertical column. Label the column with the operator function.

7. Repeat step 6 for each succeeding operator until the equation has been solved.

8. The final output column on the right side of the truth table will show the solution for the equation.

Assume now that you are to construct a truth table that will show all the possible alternative steps of a Boolean expression. The expression, shown in Figure 3-8, is $AB + B + \bar{A} = X$. Examine the expression and determine the number of inputs involved. This expression has two inputs, A and B. A column in the truth table is made for each input. The alternatives of the input are listed in binary progression order, which is 00, 01, 10, and 11. The first operator function on the left side of the expression is AB. A column is prepared for this operator. The alternatives for this operation are determined by multiplying column A by column B. The products of these alternatives are then placed in the AB column. The next operation

Given Boolean expression: AB + B + $\bar{A}$ = X

Truth table

Inputs		1st operator	2nd operator	3rd operator	Final operator
B	A	A · B	AB + B	$\bar{A}$	AB + B + $\bar{A}$
0	0	0	0	1	1
0	1	0	0	0	0
1	0	0	1	1	1
1	1	1	1	0	1

Figure 3-8. Truth table construction

is $AB + B$. The entries are found by adding column B to column AB. The response to this operation is placed in the vertical column labeled $AB + B$. The next operation involves $\bar{A}$. A column labeled $\bar{A}$ is formed. The entries for this column are inversions of column A. The final output appears in a column labeled $AB + B + A$. This function is achieved by adding the alternatives of column $AB + B$ to column A. The output of the expression is indicated by this column. It can be labeled $AB + B + \bar{A}$, or X.

Construction of this truth table permits a person to see only a few of the different problems that may arise when using this procedure. A second construction example is presented to show some different techniques. The procedure is the same but the problem is different. Figure 3-9 shows the given Boolean expression as $A+B+Cx\bar{A}+Bx\bar{C}=F$. The expression is examined and found to have three inputs. These are listed in the input columns and labeled as CBA. The alternatives are listed under these inputs in binary progression order. The first operator is listed on the left side of the expression as $A + B + C$. This operator is performed by adding the three inputs horizontally. The sums of the alternatives' are listed in the $A + B + C$ column. The next operator has an inverted input value. The A input is inverted and placed in a column labeled $\bar{A}$. The next operator is $C + B$. This operator is achieved by adding the alternatives of input B to $\bar{A}$. The resulting sums of the alternatives are listed in the $\bar{A} + B$ column. The next operator is $\bar{C}$. This is achieved by inverting input C and listing the negated values in the $\bar{C}$ column. The final step of the construction process is multiplication of the alternative values found in three columns. The columns are $A + B + C$, $\bar{A} + B$, and $\bar{C}$. The products of these three columns are listed in the final column on the right side of the table as $A + B + C$ x $\bar{A}$ +B x $\bar{C}$.

Given Boolean expression: $A + B + C \cdot \overline{A} + B \cdot \overline{C} = Y$

Inputs			1st operator	2nd operator	3rd operator	4th operator	5th operator
C	B	A	$A + B + C$	$\overline{A}$	$\overline{A} + B$	$\overline{C}$	$A + B + C \cdot \overline{A} + B \cdot \overline{C}$
0	0	0	0	1	1	1	0
0	0	1	1	0	0	1	0
0	1	0	1	1	1	1	1
0	1	1	1	0	1	1	1
1	0	0	1	1	1	0	0
1	0	1	1	0	0	0	0
1	1	0	1	1	1	0	0
1	1	1	1	0	1	0	0

Figure 3-9. Truth table construction

This represents the output of the truth table.

Conversion of a Truth Table to a Boolean Expression

A person working with digital electronics frequently needs to convert a truth table into a Boolean expression. The manufacturer of digital devices often shows component response with a truth table. Using the device generally calls for converting a truth table into a Boolean expression.

To convert a truth table to a Boolean expression, follow these operational Steps:

1. Look at the inputs and outputs of a truth table.

2. Note the output alternatives that contain a 1.

3. Refer to the inputs that produce a 1 output.

4. A 1 at the input is defined as the input letter. A 0 input is defined as a negated input letter.

5. The defined input letters that produce a 1 output are then ANDed together in a statement. (This would be something like $A*B$ or AB).

6. Write the letter statement to the right of each of the 1 outputs.

7. The lettered output statements are then ORed together to form the Boolean expression. (This would be something like $\bar{A}B + AB$).

We now go through the procedure of converting a truth table into a Boolean expression. Refer to the truth table in Figure 3-10. Notice that this is a simplified truth table that shows only the inputs and output. Look for the inputs that produce a 1. In this example, two of the four inputs have a 1 output. The inputs that produce a 1 output are 01 and 10. The other two alternatives produce a 0 output. Conversion of the binary inputs into letter expressions yields $\bar{A}B$ for 0 1 and $A\bar{B}$ for 10. The letter conversions are placed to the right of each corresponding 1 output. The resulting lettered output statements are then ORed together. The Boolean expression for this example is $\bar{A}B + A\bar{B} =F$

Truth Table Conversion of a Boolean Expression
The conversion procedure just presented deals with Boolean expressions defined as the sum of products. The primary operator of this statement is addition and the input alternatives are ANDed together. If the output of a truth table has more Os than ls, the sum of products procedure should be used to develop the Boolean expression. If the output of a truth table has more is than Os, then a different procedure must be used. This is a modification of the sum of products procedure. The resulting Boolean expression is a product of sums statement, such as $A +B+C x \bar{A}+C x \bar{A} + B + C = X$. Note that individual terms of the statement are ORed together and that the final expression is the product of the ORed terms.

Conversion of a truth table to a product of sums expression is slightly different from that of the sum of products procedure. The following steps are used to convert a truth table to a Boolean expression:

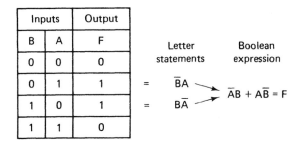

Figure 3-10. Converting a truth table to a Boolean expression

1. Look at the input and output alternatives of the truth table.

2. Note the 1 and 0 distribution of the outputs. If there is a smaller number of 0s, follow this product of sums procedure. If there is a smaller number of 1s, follow the sum of products procedure outlined previously.

3. When there is a smaller number of 0s, refer to the output location of each 0.

4. Refer to the input alternatives that produce a 0 output.

5. A 0 at one of the inputs is defined as the input letter. A 1 at the input is defined as a negated input letter.

6. The defined input letter combinations are then ORed together in a statement, such as $A + B$, or $\bar{A} + B$.

7. Write the lettered statement to the right of each of the 0 outputs.

8. The selected lettered output statements are then ANDed together to form the Boolean expression, such as $A + B \times \bar{A} + B$.

We now go through the process of converting truth table information into a Boolean expression. Refer to the truth table in Figure 3-11. Notice that the output has more is than 0s. Using the product of sums procedure, identify the location of the 0 outputs. In this truth table, only two outputs, 001 and 110, are 0. The other alternatives all produce 1 outputs. Convert the inputs that produce a 0 output into letter equivalents. A 0 produces a letter and a 1 produces a negated letter equivalent. The 001 row produces $\bar{A}\bar{B}C$ and the 110 row produces $A\bar{B}\bar{C}$. The lettered values are then ORed together. This causes 001 to be $\bar{A} + B + C$ and 110 to be $A + \bar{B} + \bar{C}$. This part of the statement should be placed to the right of the corresponding 0 output. The resulting lettered output statements are then ANDed together. The final Boolean equation is $(\bar{A} + B + C) \times (A + \bar{B} + \bar{C}) = F$. This is a typical product of sums statement.

BOOLEAN ALGEBRA PROPERTIES

Boolean algebra is used to analyze a logic circuit and to express its operation mathematically. We have used Boolean algebra to convert ex-

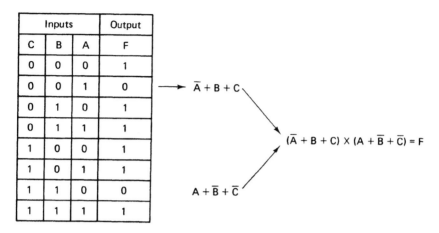

Inputs			Output
C	B	A	F
0	0	0	1
0	0	1	0
0	1	0	1
0	1	1	1
1	0	0	1
1	0	1	1
1	1	0	0
1	1	1	1

$\overline{A} + B + C$

$A + \overline{B} + \overline{C}$

$(\overline{A} + B + C) \times (A + \overline{B} + \overline{C}) = F$

Figure 3-11. Converting a truth table to a Boolean expression

pressions into truth tables, to change truth tables into mathematical expressions, and identify logic circuits from a Boolean expression. A continuation of this study leads to an investigation of Boolean theorems. A theorem is a rule or theory that has been derived from other formulas or statements. Theorems will help us to simplify logic expressions and circuits.

Single-Variable Theorems

The first group of theorems deals with single variables. In each theorem, a letter identifies the variable. The variable can have a value of either a 1 or 0. Each theorem is accompanied with a logic circuit. The circuit illustrates the theorem. The theorem is proved by substituting all possible values of the variable. Figure 3-12 shows examples of single-variable theorems.

Theorems 1 through 4 of Figure 3-12 deal with the AND function. The AND function is a multiplication operation. Any variable ANDed with 0 results in a 0 output. Likewise, a variable ANDed with a 1 is a 1 if the variable is a 1 and is 0 if the variable is 0. A negated variable ANDed on the input always produces a 0 output. Un-negated input variables produce a 1 output.

Theorems 5 through 8 of Figure 3-12 deal with the OR function. The OR function is an addition operation. In Theorem 5, a 0 added to any variable does not alter the value of the variable. Theorem 6 states that any variable ORed with a 1 produces a 1 output. Theorem 7 shows that the value of any input appears in the output when the inputs are connected

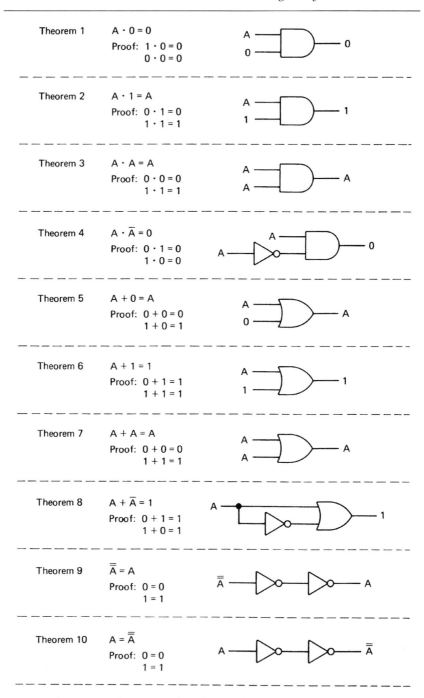

Figure 3-12. Single variable theorems

together. Theorem 8 shows that the output is always a 1 when the input variables are either negated and un-negated.

Theorems 9 and 10 of Figure 3-12 express the relationship of double negation. A variable that has been inverted twice equals its original value.

Multivariable Theorems

Boolean expressions generally deal with more than one variable. Multivariable theorems have been developed for these cases. This part of Boolean algebra is very similar to conventional algebra. The theorems are identified as the commutative, associative, and distributive theorems. Figure 3-13 shows a listing of the theorems, a logic circuit that will achieve the theorems, and a truth table that proves each one.

Theorems 11 and 12 deal with the commutative property. These theorems show that a circuit is not affected by the order or sequence of the applied variables. Theorem 11 shows the response of an AND operator and Theorem 12 refers to the OR function.

Theorems 13 and 14 refer to the response of the associative property. These theorems show that the manner in which variables are grouped has no effect on the outcome. This means that the inputs to two or more gates can be connected in any order and not alter the output. Theorem 13 shows the AND operative and 14 shows the OR operative.

Theorems 15 and 16 deal with the distributive property. These theorems show that an expression can be expanded by multiplying one term by the other term. This permits one term to be factored out of the expression. Theorem 15 shows an AND-OR operative and 16 shows an OR-AND operative.

Absorption Theorems

The absorption theorems of Figure 3-14 do not have equivalents in conventional algebra. The name *absorption* comes from the fact that one of the variables is consumed when forming a simpler equation. Three forms of the theorem are shown in the figure. In all cases, the output is a simplification of the initial expression. A truth table, equivalent circuit, and its simplification are shown with each theorem.

DeMorgan's Theorems

DeMorgan's theorems are commonly used in Boolean algebra to obtain the complement of a function or to simplify circuit equations. These

Theorem 11　Commutative property

AB = BA

Input		Output	
B	A	AB	BA
0	0	0	0
0	1	0	0
1	0	0	0
1	1	1	1

Theorem 12　Commutative property

A + B = B + A

Input		Output	
B	A	A + B	B + A
0	0	0	0
0	1	1	1
1	0	1	1
1	1	1	1

A + B = B + A

Figure 3-13. Multivariable theorems

Figure 3-13. (Continued). Multivariable theorems

Theorem 13   Associative property

A(BC) = (AB)C

C	B	A	BC	A(BC)	AB	(AB)C
0	0	0	0	0	0	0
0	0	1	0	0	0	0
0	1	0	0	0	0	0
0	1	1	0	0	1	0
1	0	0	0	0	0	0
1	0	1	1	0	0	0
1	1	0	1	0	0	0
1	1	1	1	1	1	1

A(BC) = (AB)C

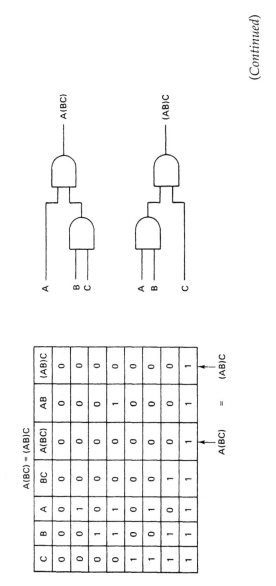

(Continued)

Figure 3-13. (Continued). Multivariable theorems

Theorem 14   Associative Property

A + (B + C) = (A + B) + C

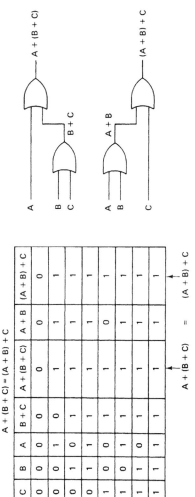

C	B	A	B + C	A + (B + C)	A + B	(A + B) + C
0	0	0	0	0	0	0
0	0	1	0	1	1	1
0	1	0	1	1	1	1
0	1	1	1	1	1	1
1	0	0	1	1	0	1
1	0	1	1	1	1	1
1	1	0	1	1	1	1
1	1	1	1	1	1	1

$$A + (B + C) = (A + B) + C$$

Theorem 15   Distributive property

A(B + C) = AB + AC

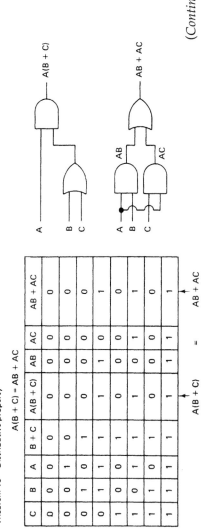

C	B	A	B + C	A(B + C)	AB	AC	AB + AC
0	0	0	0	0	0	0	0
0	0	1	0	0	0	0	0
0	1	0	1	0	0	0	0
0	1	1	1	1	1	0	1
1	0	0	1	0	0	0	0
1	0	1	1	1	0	1	1
1	1	0	1	0	0	0	0
1	1	1	1	1	1	1	1

$$A(B + C) = AB + AC$$

(Continued)

Figure 3-13. (Continued). Multivariable theorems

Theorem 16 Distributive property

$$A + BC = (A + B)(A + C)$$

C	B	A	BC	A + BC	A + B	A + C	(A + B)(A + C)
0	0	0	0	0	0	0	0
0	0	1	0	1	1	1	1
0	1	0	0	0	1	0	0
0	1	1	0	1	1	1	1
1	0	0	0	0	0	1	0
1	0	1	0	1	1	1	1
1	1	0	1	1	1	1	1
1	1	1	1	1	1	1	1

$$A + (BC) \quad = \quad AB + AC$$

or

$$B + (AC) = (A + B)(B + C)$$
$$C + (AB) = (C + A)(B + C)$$

Logic gate diagrams: inputs A, B, C producing output $A + (BC)$ (with intermediate BC); and inputs A, B, C producing output $(A + B)(A + C)$ (with intermediate terms $A + B$ and $A + C$).

Theorem 17   Absorption property

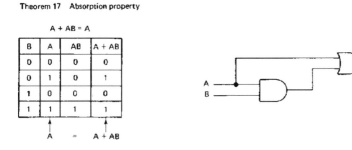

A + AB = A

B	A	AB	A + AB
0	0	0	0
0	1	0	1
1	0	0	0
1	1	1	1

A   =   A + AB

- -

Theorem 18   Absorption property

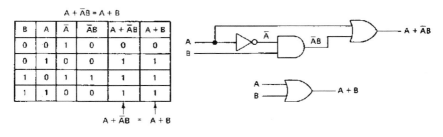

A + ĀB = A + B

B	A	Ā	ĀB	A + ĀB	A + B
0	0	1	0	0	0
0	1	0	0	1	1
1	0	1	1	1	1
1	1	0	0	1	1

A + ĀB   =   A + B

- -

Theorem 19   Absorption property

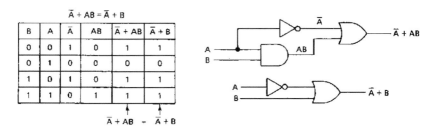

Ā + AB = Ā + B

B	A	Ā	AB	Ā + AB	Ā + B
0	0	1	0	1	1
0	1	0	0	0	0
1	0	1	0	1	1
1	1	0	1	1	1

Ā + AB   =   Ā + B

Figure 3-14. The absorption properties

theorems are extremely useful in simplifying expressions in which a
product or sum of variables is inverted. Two versions of the theorems are
shown in Figure 3-15. Theorem 20 shows that when the OR sum of two
variables is negated, it is the same as negating individual variables and
then ANDing them together. Theorem 21 shows that when the AND product of two variables is negated, it is the same as negating each variable
individually and then ORing them together. These theorems are proven
with a truth table.

The dual or equivalent of a Boolean expression can be found by
employing DeMorgan's theorem. It is implemented through the follow-

Theorem 20 DeMorgan's theorem (form A)

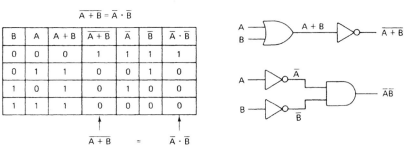

$$\overline{A + B} = \overline{A} \cdot \overline{B}$$

B	A	A + B	$\overline{A + B}$	$\overline{A}$	$\overline{B}$	$\overline{A} \cdot \overline{B}$
0	0	0	1	1	1	1
0	1	1	0	0	1	0
1	0	1	0	1	0	0
1	1	1	0	0	0	0

$$\overline{A + B} \quad = \quad \overline{A} \cdot \overline{B}$$

Theorem 21 DeMorgan's theorem (form B)

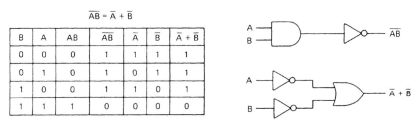

$$\overline{AB} = \overline{A} + \overline{B}$$

B	A	AB	$\overline{AB}$	$\overline{A}$	$\overline{B}$	$\overline{A} + \overline{B}$
0	0	0	1	1	1	1
0	1	0	1	0	1	1
1	0	0	1	1	0	1
1	1	1	0	0	0	0

Figure 3-15. DeMorgan's theorems

ing three steps.

1. Replace all the OR operators (+) with AND operators (x) or the AND operators (X) with an OR operator (+).

2. Replace all variables with their complements (A with $\overline{A}$, and B with $\overline{B}$).

3. Complement the entire expression $(A + B)$ with $(\overline{A + B})$.

An example of the DeMorgan process is shown in Figure 3-16. The equation is shown in its original form. The conversion steps are then performed. Modification of the equation is shown. The resulting equation is then shown. Logic circuits and corresponding truth tables are given to show equivalency.

Another example of the DeMorgan process is shown in Figure 3-17. This example has three variables. The procedure is outlined in the same manner. Examine the equation, review the conversion steps, and determine if the resulting circuit is an equivalent of the original equation.

Original equation: $\overline{A} + \overline{B} = X$

Step 1: Replace OR operators with AND operators.

$$\overline{A} + \overline{B} = X \longrightarrow \overline{A} \cdot \overline{B} = X$$

Step 2: Complement each variable.

$$\overline{A} \cdot \overline{B} = X \longrightarrow A \cdot B = X$$

Step 3: Complement the entire expression.

$$A \cdot B = X \longrightarrow \overline{A \cdot B} = X$$

Solution: $\overline{A} + \overline{B} = \overline{A \cdot B}$

Logic circuit equivalents:

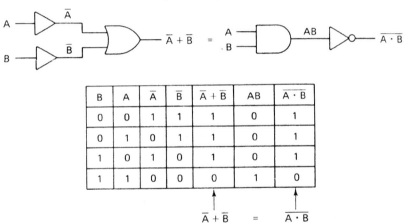

B	A	$\overline{A}$	$\overline{B}$	$\overline{A} + \overline{B}$	AB	$\overline{A \cdot B}$
0	0	1	1	1	0	1
0	1	0	1	1	0	1
1	0	1	0	1	0	1
1	1	0	0	0	1	0

$$\overline{A} + \overline{B} \quad = \quad \overline{A \cdot B}$$

Figure 3-16. Applying DeMorgan's theorem

SIMPLIFYING EQUATIONS USING BOOLEAN ALGEBRA

One of the chief uses of Boolean algebra is in the simplification of a logic expression. A simplified circuit involves fewer gates and is less expensive to construct. This process entails selecting a theorem and placing it in an equation so that it will reduce the original expression. Several theorems may be used to accomplish the end result. A few examples are presented here to illustrate the procedure. As a rule, it takes a great deal of practice in this procedure to become proficient.

Assume now that you are given a Boolean expression and its equivalent logic circuit. Simplify the expression and prove that the reduced cir-

Original equation: $\overline{A} \cdot \overline{B} \cdot \overline{C} = X$.

Step 1: Change operators.

$$\overline{A} + \overline{B} + \overline{C} = X$$

Step 2: Complement each variable.

$$\overline{\overline{A}} + \overline{\overline{B}} + \overline{\overline{C}} = X$$

Since each variable is now double negated, the negation is canceled.

$$A + B + C = X$$

Step 3: Complement the entire equation.

$$\overline{A + B + C} = X$$

Solution: $\overline{A} \cdot \overline{B} \cdot \overline{C} = \overline{A + B + C}$

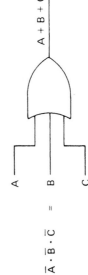

Figure 3-17. Applying DeMorgan's theorem

Given Boolean expression: A(A + B) = X

or

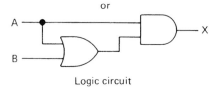

Logic circuit

Step 1: Expand using Theorem 15 (distributive).
Multiply A through (A + B).

AA + AB = X

Step 2: Theorem 3 shows that A · A = A.
The expression now becomes:

A + AB = X

Step 3: Theorem 17 (absorption) shows that A + AB = A.
The simplified expression then becomes:

A = X

Step 4: The reduced logic circuit is simply a wire connected
between the A input and the X output.

Step 5: The truth table shows that A and A(A + B) are equal.

B	A	A + B	A(A + B)
0	0	0	0
0	1	1	1
1	0	1	0
1	1	1	1

A   =   A(A + B)

Figure 3-18. Example 1: Simplifying a Boolean equation

cuit is the equivalent of the original statement. Figure 3-20 shows the simplification procedure for Example 1. Follow the outlined procedure for
reducing this expression.

Example 2 is shown in Figure 3-19. This equation involves a different procedure. There is no set procedure for the simplification process. As
a rule, the equation dictates the procedure to be followed in its simplifica-

Given Boolean expression: $(\overline{A} + B)(A + B) = Q$

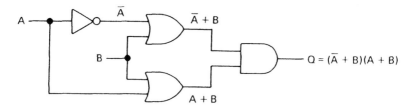

Step 1: Expand the equation using Theorem 15 (distributive).
Multiply $\overline{A} + B$ through $A + B$. The expression then becomes

$$A\overline{A} + \overline{A}B + AB + BB = Q$$

Step 2: Theorem 3 shows that $A \cdot A = A$. Therefore,

$$A\overline{A} + \overline{A}B + AB + B = Q$$

Step 3: Theorem 4 shows that $A \cdot \overline{A} = 0$. Thus

$$0 + \overline{A}B + AB + B = Q$$

Step 4: Theorem 4 shows that $A + 0 = A$. The equation becomes

$$\overline{A}B + AB + B = Q$$

Step 5: Theorem 15 permits B to be factored out of the expression. The equation then becomes

$$B(\overline{A} + A + 1) = Q$$

Step 6: Theorem 8 shows $A + \overline{A} = 1$. The expression becomes

$$B(1 + 1) = Q$$

Step 7: Since $1 + 1 = 1$, the expression becomes

$$B(1) = Q$$

Step 8: Theorem 2 shows that $A \cdot 1 = A$. The expression then becomes

$$B = Q$$

Figure 3-19. Example 2: Simplifying a Boolean equation

(Continued)

Step 9: The reduced circuit shows that a wire connecting the output to
input B would accomplish the same thing as the original statement.

Step 10: The truth table shows that B = Q.

					Output
B	A	$\overline{A}$	$\overline{A}$ + B	A + B	($\overline{A}$ + B)(A + B)
0	0	1	1	0	0
0	1	0	0	1	0
1	0	1	1	1	1
1	1	0	1	1	1

B   =   Q

Figure 3-19 (*Continued*). Example 2: Simplifying a Boolean equation

tion. Refer to Figure 3-19 and follow the outlined procedure for this simplification process.

Example 3 is shown in Figure 3-20. This equation has three variables.
It is somewhat more complex than the previous examples. We use the procedure outlined in the previous examples as a guide for simplifying this
equation.

UNIVERSAL BUILDING BLOCKS

Given a Boolean expression, we can build a logic circuit that will accomplish its operation. In the same manner, given a logic circuit, we can
develop a Boolean expression that will show its operation. As an example,
the Boolean expression $(A + B)\overline{C}$ represents a logic circuit in which A is
ORed to B and then ANDed with NOT C. Figure 3-21 shows a logic circuit
that will accomplish this expression.

Up to this point in the study of logic gates we have used AND, OR,
and NOT gates to achieve logic functions. These Boolean operators are
considered to be the basic building blocks of a logic system. They can be
used in combinations to accomplish all logic functions.

NAND and NOR gates have a very interesting property that needs
to be explored more fully. These two gates can be used to build AND, OR,
or NOT gates. Because of this, NAND or NOR gates are all that we need to

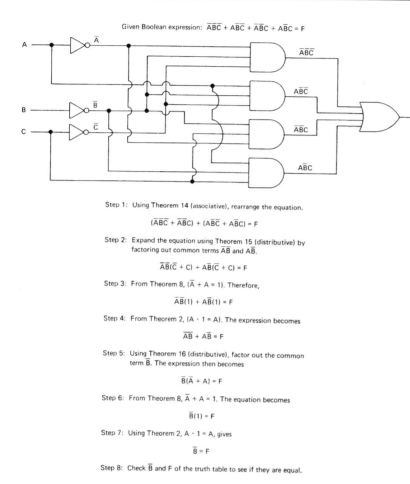

Step 1: Using Theorem 14 (associative), rearrange the equation.

$$(\overline{A}\overline{B}\overline{C} + \overline{A}\overline{B}C) + (A\overline{B}\overline{C} + A\overline{B}C) = F$$

Step 2: Expand the equation using Theorem 15 (distributive) by factoring out common terms $\overline{A}\overline{B}$ and $A\overline{B}$.

$$\overline{A}\overline{B}(\overline{C} + C) + A\overline{B}(\overline{C} + C) = F$$

Step 3: From Theorem 8, $(\overline{A} + A = 1)$. Therefore,

$$\overline{A}\overline{B}(1) + A\overline{B}(1) = F$$

Step 4: From Theorem 2, $(A \cdot 1 = A)$. The expression becomes

$$\overline{A}\overline{B} + A\overline{B} = F$$

Step 5: Using Theorem 16 (distributive), factor out the common term $\overline{B}$. The expression then becomes

$$\overline{B}(\overline{A} + A) = F$$

Step 6: From Theorem 8, $\overline{A} + A = 1$. The equation becomes

$$\overline{B}(1) = F$$

Step 7: Using Theorem 2, $A \cdot 1 = A$, gives

$$\overline{B} = F$$

Step 8: Check $\overline{B}$ and F of the truth table to see if they are equal.

Figure 3-20. Example 3: Simplifying a Boolean expression

build a logic system. NAND and NOR gates are considered to be universal building blocks. The logic symbol, truth table, and Boolean equivalent of these gates are shown in Figure 3-22.

The NAND Gate

In looking through manufacturers' literature, you will find that the NAND gate is more widely used than any other type of gate. Because of this, we will look at the NAND gate to see how it is used to make other types of gates. The NAND gate is truly a universal building-block gate. A NOT gate can be made from a NAND gate by simply connecting all inputs

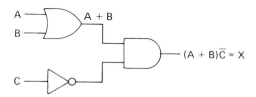

Figure 3-21. Logic symbol, truth table, and Boolean equation of NAND and NOR gates

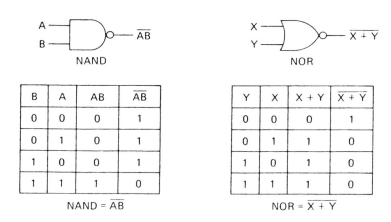

Figure 3-22. Logic symbol, truth table, and Boolean equation of NAND and NOR gates.

together, as shown in Figure 3-23. In this case, if input A is 1, the output will be 0, or $\bar{A}$. In the same manner, if the input is negated ($\bar{A}$), or 0, the output will be 1, or A. The truth table of Figure 3-22 shows that the output of a two-input NAND gate is ($\overline{AB}$). This means that when $A = 0$ and $B = 0$, the output ($\overline{AB}$) is 1. In the same manner, when $A = 1$ and $B = 1$, the output ($\overline{AB}$) is 0. We simply connect the inputs of a NAND gate together to accomplish this operation. It should be understood that when the inputs of a NAND gate are tied together, the gate will achieve a NOT function.

A NAND gate can be used to produce an AND function by connecting the output of one gate to the input of a NOT gate. This configuration requires the use of two NAND gates. Figure 3-24 shows NAND gates connected to accomplish the AND function. Expressed in Boolean form, the first NAND gate will accomplish ($\overline{AB}$). The second NAND gate, being used as a NOT gate, will cause the expression to be ($\overline{AB}$)', or AB. The two-NAND-gate combination therefore produces the AND function. The input

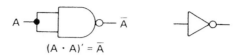

$(A \cdot A)' = \overline{A}$

Figure 3-23. A NAND gate achieving a NOT gate

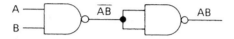

Figure 3-24. NAND-gate to AND-gate conversion

is *A* and *B*, and the output is *AB*. It should be understood that when the output of one NAND gate is connected to a NOT gate, the configuration will accomplish the AND function.

NAND gates can be used to produce an OR function. Figure 3-25 shows how NAND gates are connected to achieve the OR function. Each input is applied to a NOT gate. The NOT function is achieved by a NAND gate with its inputs connected together. The *A* and *B* inputs then become $\overline{A}$ and $\overline{B}$. This is applied to a two-input NAND gate. The output of this gate then becomes $(\overline{AB})$. DeMorgan's law must be applied to the expres-

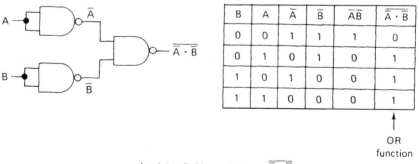

B	A	$\overline{A}$	$\overline{B}$	$\overline{A}\overline{B}$	$\overline{\overline{A} \cdot \overline{B}}$
0	0	1	1	1	0
0	1	0	1	0	1
1	0	1	0	0	1
1	1	0	0	0	1

OR
function

Applying DeMorgan's law to $\overline{\overline{A} \cdot \overline{B}}$:

Step 1: $\overline{\overline{A} + \overline{B}}$

Step 2: $\overline{\overline{A}} + \overline{\overline{B}}$ or $\overline{\overline{A} + B}$

Step 3: $\overline{\overline{A + B}}$ or $A + B$

Figure 3-25. NAND-gate to OR-gate conversion

sion for it to be meaningful. To apply DeMorgan's law, change the sign of the operator, complement each variable, and then complement the entire expression. This will produce an output of $A + B$, which is the OR function. Figure 3-25 shows the steps needed to apply DeMorgan's law to the output of the gate circuit. Since the output of this gate circuit is $A + B$, we have shown that NAND gates can be used to achieve the OR function.

The NOR Gate

The NOR gate is considered to be a universal building block. It is used as the construction basis of some IC families. Because of this, we will look at the operation of this gate to see how it is used to achieve other gate functions. The gate circuit connections are similar to those of the NAND gate.

A NOT gate can be made from a NOR gate by simply connecting all the inputs together. This circuit connection permits the NOR gate to respond as an inverter. The truth table of a NOR gate and its logic symbol is shown in Figure 3-26. The truth table shows that when the inputs are all 0s, the output will be 1, and when the inputs are all 1s, the output will be 0. This means that inputs connected together will produce inversion, or negation. Figure 3-26 shows a NOR gate connected to achieve the NOT function. With the inputs connected together, the gate will have A applied to both inputs at the same time. The output is $(\overline{A + A})$, which can be simplified by theorems to $\overline{A}$. This means that when the inputs of a NOR gate are tied together, the output will be a complement, or inversion, of the input.

NOR gates can be used to produce an AND function. Figure 3-27 shows the logic gate connections needed to construct an AND gate from NOR gates. Each input is applied to a separate NOT gate. The NOT function is achieved by connecting the inputs of a NOR gate together. The A and B inputs then become A and B. This is then applied to a two-input NOR gate. The output of this gate is $(\overline{A + B})$. DeMorgan's law must be applied. Change the sign of the operator, complement each variable, and then complement the entire expression. This causes $(\overline{\overline{A} + \overline{B}})$ to be AB, which is the AND function. Figure 3-27 shows the DeMorgan operation and truth table of the logic circuit. Since the output of this configuration is AB, it shows that NOR gates can be used to accomplish the AND function.

NOR gates can be used to produce an OR function by connecting the output of one gate to the input of a NOT gate. This configuration requires the use of two NOR gates. Figure 3-28 shows an OR gate produced by NOR gates. Expressed in Boolean form, the first NOR gate will

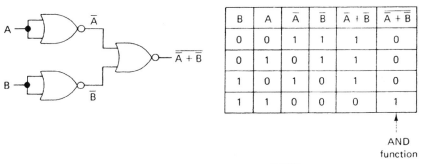

NOT gate circuit   NOT symbol

Figure 3-26. NAND-gate to NOT-gate conversion

B	A	$\bar{A}$	$\bar{B}$	$\bar{A} + \bar{B}$	$\overline{\bar{A} + \bar{B}}$
0	0	1	1	1	0
0	1	0	1	1	0
1	0	1	0	1	0
1	1	0	0	0	1

AND
function

Applying DeMorgan's law to $\overline{\bar{A} + \bar{B}}$:

Step 1: $\overline{\bar{A} \cdot \bar{B}}$

Step 2: $\overline{\bar{A}} \cdot \overline{\bar{B}}$ or $\overline{A \cdot B}$

Step 3: $\overline{\bar{A} \cdot \bar{B}}$ or $A \cdot B$ (AND function)

Figure 3-27. NOR-gate to AND-gate conversion

accomplish $(\overline{A + B})$. The second NOR gate, being used as a NOT gate, will cause the expression to be $(\overline{\bar{A} + \bar{B}})$, or $A + B$. The two-NOR-gate configuration therefore produces the OR function. The input is A and B, and the output is $A + B$. It should be understood that when the output of one NOR gate is connected to a NOT gate, the logic circuit will accomplish the OR function.

KARNAUGH MAPS

   In 1953 Maurice Karnaugh of Bell Laboratories published an article explaining how to simplify Boolean expressions with a special truth table. This truth table, sometimes called a map, is a graphic method of simplifying a Boolean expression. Today, this simplification process is called a Karnaugh map. Karnaugh maps represent an alternate method of Boolean

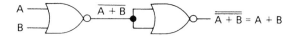

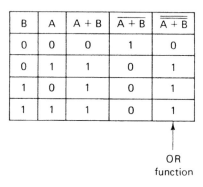

B	A	A + B	$\overline{A + B}$	$\overline{\overline{A + B}}$
0	0	0	1	0
0	1	1	0	1
1	0	1	0	1
1	1	1	0	1

OR
function

Figure 3-28. NOR-gate to OR-gate conversion

expression reduction that does not use theorems. For expressions with 2, 3, or 4 variables, this method is easy to follow and can be performed faster than the theorem procedure. Expressions containing more than 4 variables become rather complex and somewhat more difficult to manage. Karnaugh mapping is generally preferred over the theorem method of simplification because it can be used quickly and conveniently.

Two-Variable Karnaugh Maps

A Karnaugh map is a modified version of a conventional truth table. It translates truth table information into a graphic display. This display shows the relationship between logic inputs and the desired output. Figure 3-29 shows a truth table and a Karnaugh map for a conventional two-variable Boolean expression. With 2 inputs, there must be 2^2, or 4, blocks to display the different combinations. The inputs or variables are identified by letters assigned to specific blocks. The upper left square, or block 1, identifies the function of $\bar{A}$ and $\bar{B}$. The upper right cell, or block 2, shows the operation of $\bar{A}B$. The lower left block shows $A\bar{B}$. Block 4 shows the function of AB. Note that the squares are identified by horizontal rows labeled $\bar{A}$ and A with vertical columns identified by the letters $\bar{B}$ and B. The output or product term of a specific input combination is placed in a corresponding block. Outputs are generally described as minterms. The minterms of a Boolean expression appear in blocks according to the place-

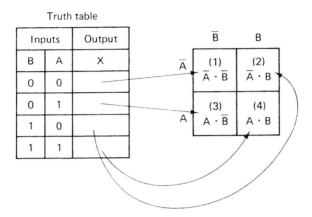

Truth table

Inputs		Output
B	A	X
0	0	
0	1	
1	0	
1	1	

	$\overline{B}$	B
$\overline{A}$	(1) $\overline{A} \cdot \overline{B}$	(2) $\overline{A} \cdot B$
A	(3) $A \cdot \overline{B}$	(4) $A \cdot B$

Figure 3-29. Two-variable Karnaugh map and truth table

ment of the inputs. In the map of Figure 3-29 minterms are placed in the corresponding blocks.

Let us map the Boolean expression of Figure 3-30 to show the use of a Karnaugh map. This Boolean expression has two variables. The map must have 4 blocks to accommodate the inputs. These are located on the map as indicated. A "1" is now placed in the squares of the map, which represent the product terms of the expression. Note the connecting arrows showing the placement of the product terms. The Boolean expression is now displayed on the two-variable Karnaugh map as ls. These represent the product terms of the expression.

When the blocks of a Karnaugh map have been filled in, it can be examined to see if the expression can be simplified. If is appear in adjacent blocks the expression can be simplified. If the map does not possess adjacent ls, the expression cannot be simplified. In our example, adjacent is appear at two positions of the map. These is are generally looped together

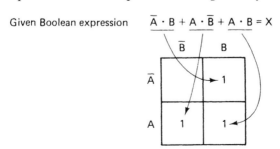

Given Boolean expression $\quad \overline{A} \cdot B + A \cdot \overline{B} + A \cdot B = X$

	$\overline{B}$	B
$\overline{A}$		1
A	1	1

Figure 3-30. Boolean expression conversion to a Karnaugh map

to show their relationship. Figure 3-31 shows the is looped together on the map. These loops mean that we will have two terms that are ORed together in the simplified expression.

Let us now take a look at the looped terms of the Karnaugh map of Figure 3-31. Loop 1 shows that B occurs in both blocks along with $\bar{A}$ and A. This means that the terms $\bar{A}$ and A can be eliminated according to the rules of Boolean algebra. Loop 1 represents B in the simplified expression. Loop 2 shows that A occurs in both blocks along with $\bar{B}$ and B. This means that the $\bar{B}$ and B terms can be canceled. This leaves A represented by loop 2. The simplified expression for this map is $A + B = X$.

The first opportunity to simplify a Boolean expression with a Karnaugh map generally tends to be a little confusing. Actually, the procedure is rather easy to follow when it is performed in operational steps. These steps are summarized below.

1. Divide the original Boolean expression into product terms or minterms.

2. Identify the number of variables in the expression.

3. Prepare a block map and identify the location of each input variable.

4. Record is on the map structure for each product term of the expression.

5. Loop adjacent squares if possible.

6. Simplify the looped squares by dropping the complemented terms.

Figure 3-31. Adjacent terms looped together in a Karnaugh map

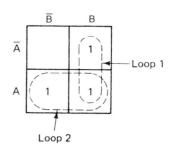

7.   OR the remaining terms.

8.   Write the simplified Boolean expression.

Three-Variable Karnaugh Maps

Now consider a Karnaugh map that has a three-variable input. This map must have 2^3, or 8, blocks in its structure. Figure 3-32 shows the structure of a three-input map. The inputs are identified by the letter designations of A, B, and C. Keep in mind that a letter such as A, B, or C represents a 1 and the negated version of $\bar{A}$, $\bar{B}$, or $\bar{C}$ is 0. Individual blocks of the map show the location of representative product terms of the input variables. Note that the letter designations $\bar{B}$ and B are each assigned a horizontal row. These are located on the left side of the map. The letters $\bar{A}$ and A are assigned to vertical rows on the top of the map. Notice that $\bar{A}$ and A are each centered over a line that divides two columns. This shows that the designated letter $\bar{A}$ or A applies to the columns on either side of the line. The letters $\bar{C}$ and C are assigned to vertical columns at the bottom of the map. $\bar{C}$ is assigned to the right- and left-most columns and C is assigned to the two center columns. Note where the indicated product terms appear in the blocks of the map. See if you agree with the placement of the minterms.

To show the use of a Karnaugh map with a three-variable input, consider the Boolean expression $\bar{A}\bar{B}\bar{C} + \bar{A}\bar{B}C + A\bar{B}C + A\bar{B}\bar{C} = X$. This expression along with a representative three-variable Karnaugh map is shown in Figure 3-33. The map has 8 squares to accommodate the three-input variables. Ones are placed on the appropriate squares of the map to represent the product terms of the expression. Arrows are used to show the placement of the product terms in the appropriate blocks. The map is reproduced in part (c) of Figure 3-33, which shows where adjacent terms

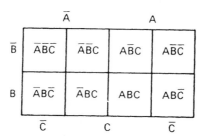

Figure 3-32. Three-variable Karnaugh map

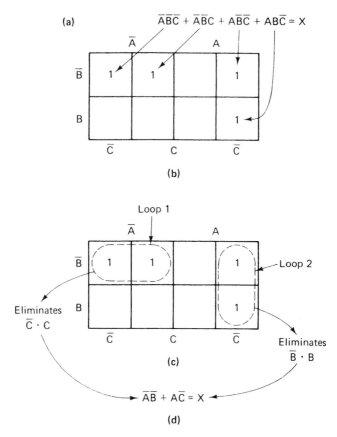

Figure 3-33. Karnaugh map use: (a) Boolean expression; (b) Karnaugh map layout; (c) Karnaugh map looping; (d) simplified Boolean expression

are connected together with loops. Loop 1 is used to eliminate $\bar{C}$ and C of the two product terms while loop 2 eliminates $\bar{B}$ and B. The simplified Boolean expression then becomes $\bar{A}\bar{B} + A\bar{C}$. The procedure steps for a three-variable Karnaugh map are the same as those identified in the two-variable procedure.

Four-Variable Karnaugh Maps

A four-variable Karnaugh map is shown in Figure 3-34. Each square of this map represents one of the four-variable product terms or minterms. This map must have 16 squares to accommodate the 4 input variables. Note the location of the input variables on the map. $\bar{A}$ and A are centered

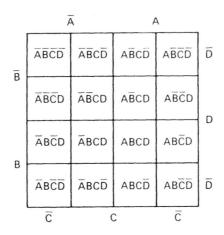

Figure 3-34. Four-variable Karnaugh map

on vertical lines at the top of the map. $\bar{B}$ and B are centered on horizontal lines on the left side of the map. $\bar{C}$ is assigned to the left- and right-most vertical columns at the bottom, with C on the center line. $\bar{D}$ is assigned to the top-and bottom-most horizontal rows on the right side with D on the center horizontal line. Keep in mind that a centered letter applies to the columns or rows on each side of the line. The resulting minterms of this map appear in the designated blocks. See if you agree with the placement of the designated minterms.

To show the use of a Karnaugh map in simplifying a four-variable Boolean equation, consider the expression $\bar{A}B\bar{C}D + \bar{A}\bar{B}\bar{C}D + \bar{A}BCD + \bar{A}BCD + A\bar{B}\bar{C}\bar{D} + A\bar{B}\bar{C}D = X$. This expression, along with a four-input Karnaugh map is shown in Figure 3-35. Ones are placed in blocks to represent different product terms of the Boolean expression. Arrows are used to show the placement of the product terms. The map is then reproduced in part (c) of Figure 3-35, which shows where adjacent terms are connected together with loops. Loops 1, 2, 3, and 4 combine to eliminate the $\bar{B}$ B and $\bar{C}$ C complements. The remaining terms are combined to form $\bar{A}D$. Loop 5 eliminates the D^* D complement. The remaining terms are combined to form $A\bar{B}\bar{C}$. ORing these terms makes the simplified expression become $\bar{A}D + A\bar{B}\bar{C}$. The general procedural steps for this map are the same as those of the two- and three-variable map development.

(a) $\overline{A}B\overline{C}D + \overline{A}B\overline{C}\overline{D} + \overline{A}\overline{B}CD + \overline{A}BCD + A\overline{B}\overline{C}\overline{D} + A\overline{B}\overline{C}D = X$

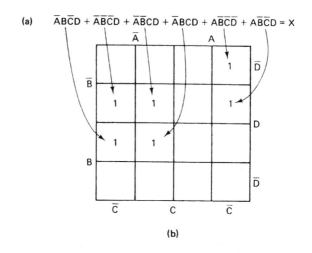

(b)

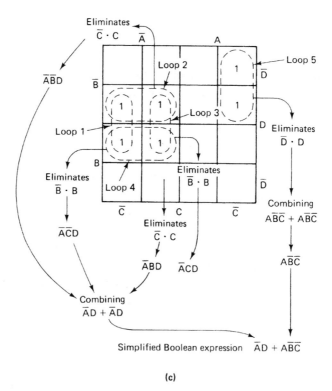

(c)

Figure 3-35. Four-variable Karnaugh map development: (a) Boolean expression; (b) four-variable Karnaugh map; (c) looping.

More Karnaugh Map Considerations

Some variations of the Karnaugh map are presented here for your consideration. As a rule, these considerations deal with the location of variables and unusual looping procedures.

Some Karnaugh map makers combine two input variables for placement on vertical columns and horizontal rows. Figure 3-36 shows a map laid out in this manner. This procedure produces the same results as the individual letter column and row input assignments discussed previously. In fact, the same block assignments can be made by pulling out one letter from the two letter combination. Note the single letter assignments for the map. In this case, the location of $\bar{D}$ and D are on the bottom of the map with $\bar{B}$ and B on the right side. The functional operation of a map structured in this manner accomplishes the same operation. Letter placement can be oriented in any position vertically, horizontally, top, bottom, right, or left and still accomplish the same functional operation. We prefer to use the single letter assignments for the blocks of a Karnaugh map.

The looping of adjacent terms on a Karnaugh map can lead to some unusual considerations. Refer to the Karnaugh map of Figure 3-37 that has

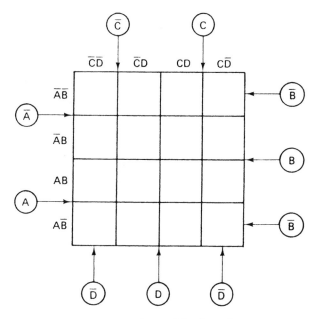

Figure 3-36. Two-letter block assignment

wraparound looping. The Boolean expression represented by this map is $\bar{A}B\bar{C}\bar{D} + AB\bar{C}\bar{D} + \bar{A}BCD + ABC\bar{D} = X$. Loops 1 and 2 connect adjacent blocks located on the left and right sides of the map. This can be done when the map is viewed as a cylinder. Loops 3 and 4 show adjacent blocks connected in the vertical columns. Loops 1 and 2 cancel the $\bar{C}\,C$ complements and loops 3 and 4 cancel the $\bar{A}\,A$ complements. The remaining variables of B and D are combined to form the simplified expression of $B\bar{D} = X$.

Another unusual looping problem is shown in Figure 3-38. The Boolean expression represented by this map is $\bar{A}\bar{B}\bar{C}\bar{D} + \bar{A}B\bar{C}D + A\bar{B}\bar{C}\bar{D} + A\bar{B}\bar{C}D = X$. Loops 1 and 2 connect adjacent blocks located in the top and

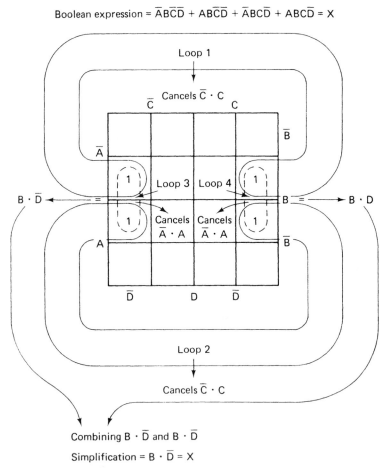

Figure 3-37. Karnaugh map side looping

bottom horizontal rows. Loops 3 and 4 connect adjacent blocks located on the top and bottom of the vertical columns. The vertical columns are viewed as a connected cylinder structure. Loops 1 and 2 cancel the $\bar{D}$ D complements and loops 3 and 4 cancel the $\bar{A}A$ complements. The remaining variables of $\bar{B}\bar{C}$ are combined to form the simplified expression of $\bar{B}\bar{C} = X$.

Figure 3-39 shows a Karnaugh map with corner looping. The Boolean expression represented by this map is $\bar{A}\bar{B}\bar{C}\bar{D} + \bar{A}\bar{B}C\bar{D} + A\bar{B}\bar{C}\bar{D} + A\bar{B}C\bar{D} =$

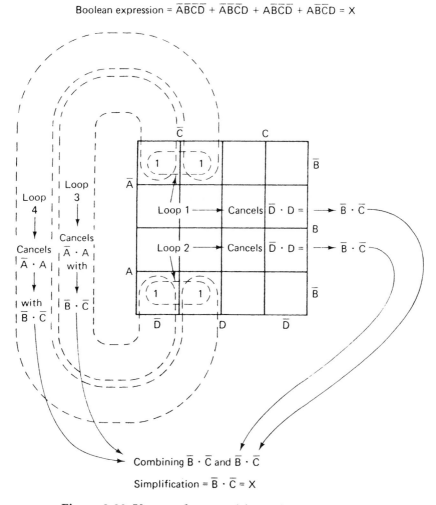

Boolean expression = $\overline{A}\,\overline{B}\,\overline{C}\,\overline{D}$ + $\overline{A}\,\overline{B}\,\overline{C}D$ + $A\,\overline{B}\,\overline{C}\,\overline{D}$ + $A\,\overline{B}\,\overline{C}D$ = X

Combining $\bar{B} \cdot \bar{C}$ and $\bar{B} \cdot \bar{C}$

Simplification = $\bar{B} \cdot \bar{C} = X$

Figure 3-38. Karnaugh map with top-bottom looping

X. Loops 1 and 3 connect adjacent blocks at the top and bottom corners. Loops 2 and 4 connect adjacent blocks at the top and bottom of the left and right side of the map. The map can be viewed as a ball with all corners connected together in a composite structure. Loops 1 and 3 cancel the $\bar{C} C$ complements while loops 2 and 4 cancel the $\bar{A} A$ complements. The remaining variables of $\bar{B}\bar{D}$ are combined to form the simplified expression of $\bar{B}\bar{D} = X$.

Boolean expression = $\overline{A}\,\overline{B}\,\overline{C}\,\overline{D} + \overline{A}\,\overline{B}\,C\,\overline{D} + A\,\overline{B}\,\overline{C}\,\overline{D} + A\,\overline{B}\,C\,\overline{D} = X$

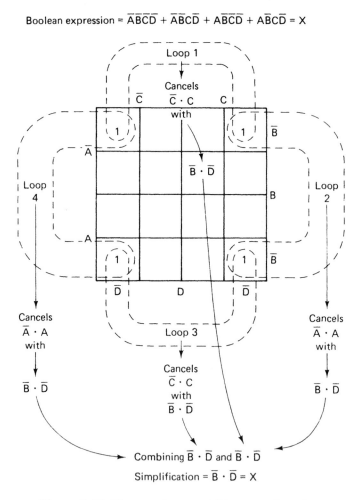

Figure 3-39. Karnaugh map with corner looping

SUMMARY

One important application of Boolean algebra is to express how logic functions are achieved mathematically. A Boolean expression is used to show how an equation responds with respect to its input and output. The inputs and outputs are identified by designated letters. A Boolean expression can be described by a mathematical statement, a logic circuit, or a truth table.

When one logic gate is used to accomplish a function, it is considered to be a single-level operator. Two gates connected so that the output of one gate feeds the input of the second gate is defined as a two-level function. Gate circuit functions achieved by two-, three-, and four-level operators occur quite frequently in digital systems.

The process of changing a logic circuit to a Boolean expression has several operational steps. These include: (1) identification of the inputs, (2) identification of the operator, (3) partial statement formation of the input-operator function, (4) partial statement formation of the other first-level inputs, (5) identification of the second-level function, (6) labeling the second-level function and identification of its operator, and (7) output-statement formation.

The process of changing a Boolean expression into a logic circuit has several operational steps. These include: (1) identification of the input and the output of the statement, (2) primary operator identification, (3) gate selection that will achieve the operator, (4) input and output label assignments, (5) input-operator identification, (6) gate selection for the operator, (7) labeling the secondary inputs, (8) additional operator assignments if needed, and (9) connection of the secondary inputs to common points.

A truth table is a graphic display of all possible combinations of the input and output variables of a Boolean equation. The construction of a truth table includes: (1) inspecting the Boolean statement, (2) making a column for each input, (3) starting with the first operator function on the left side of the expression, (4) performing the operator function, (5) preparing an output column, (6) solving the next operation, (7) repeating the procedure for the next operator, and (8) recording the final output in the right-hand column.

Conversion of a truth table to a Boolean expression is a common practice. This is achieved by: (1) identifying the truth table inputs and outputs, (2) selecting the outputs that contain a 1, (3) locating the inputs that produce a 1 output, (4) identifying the inputs with a 1 as a letter and a 0 input as a negated letter, (5) ANDing the input letters together in a statement, (6) placing a lettered statement to the right of each 1 output, and (7) ORing the lettered statements together to form a sum of products Boolean expression.

Converting a truth table to a Boolean expression when the outputs are nearly all 1s calls for a modification of the procedure. The end result is a product of sums statement. The procedure steps include: (1) inspecting the input and output of the truth table, (2) referring to the 0 outputs of the truth table when there is a small number of 0s, (3) defining a 0 at the input as a letter and a 1 as a negated letter, (4) ORing the defined input letters together, (5) placing the lettered statement to the right of each of the 0 outputs, (6) ANDing the identified statements together to form a product of sums Boolean expression.

AND, OR, and NOT gates are considered to be the basic building blocks of a logic system. NAND and NOR gates are considered to be universal building blocks. NAND and NOR gates are all we need to build a logic system.

A Karnaugh map is a modified version of a conventional truth table. It translates truth table information into a graphic display. The inputs of a Karnaugh map are identified by letters assigned to specific blocks. A two-input map must have four blocks. To convert a Boolean expression into a Karnaugh map, divide the expression into minterms, then identify the variables. Prepare a map that identifies each variable. Record 1s on the map for each product term. Loop adjacent squares. Simplify by dropping the complemented terms. OR the remaining terms.

A three-variable Karnaugh map has 8 squares or blocks. A four-variable map has 16 blocks to accommodate the inputs. The procedural steps for 3 and 4 variable maps are the same as those of the two input map. With 3 and 4 input maps, the looping procedure is somewhat unusual. Adjacent squares may appear on the top and bottom, the sides, or the corners. The map is generally viewed as a cylinder, a tube, or a ball when connecting adjacent squares.

Chapter 4

Combinational Logic Gates

INTRODUCTION

One of the most important characteristics that a digital system must possess is the ability to communicate with an operator or another system. Communication is achieved through a series of codes, or languages. A digital system manipulates these signals in its operation. The system must respond to these signals and produce an output that is indicative of the applied data. As a rule, the system must accept an input signal, which may be in the form of a spoken word, a keystroke from a keyboard, or a decimal number from a calculator, and convert it to a usable code. Internally, the system may respond to binary, octal, or hexadecimal data. The initial operational step is called *encoding*. The system then manipulates data according to instructions. Data are then distributed to different parts or elements of the system. This is the transmission function. Data transmission may involve the use of multiplexers and demultiplexers. The output of the system is responsible for converting digital information back into "real-world" codes or languages that can be identified by the operator or different pieces of equipment. This is the *decoding* function. Decoder information is generally transmitted to a display or output device. A digital system is responsible for data conversion such as encoding and decoding and must be capable of transmitting data internally to different parts and externally to different units of the system. Comparators are used for determining whether a given number is larger than, smaller than or equal to another number. Data conversion and transmission are very important system functions.

ENCODERS

Encoders are used to convert one or more input data lines into multi-line output data. A typical encoder converts decimal data into BCD data. The keyboard of a calculator is an example of a decimal input. Each key triggers a switch. The switch, or push button, represents a distinct decimal value. It simply actuates a combinational logic input circuit. This circuit then produces an output that is representative of the input number value. The output is usually a coded value of the input. Typical encoder outputs are in the form of a BCD value. In a sense, one active input signal produces a code output that has many active signals.

To demonstrate the operation of an encoder, let us assume that we have energized an input identified as the decimal number 5. When we do this the circuit will create a code that is representative of the binary value of the decimal number 5. The output of the encoder will be 0101_2. When an input line is activated, it triggers the logic circuitry so that the output will be $D = 0$, $C = 1$, $B = 0$, and $A = 1$. The output is a BCD representation of the decimal input.

Figure 4-1 shows one way of producing a 0101 BCD value for the decimal number 5. When the input is energized it feeds an active high value to the C and A output lines. Note that each output line is attached to a pull-up resistor. This causes the output to be low when there is no input voltage present. When the number 5 is energized it causes the C-A lines to be high and the D-B lines to be low. The output is 0101, which is the BCD equivalent of the decimal number 5. This is considered to be an active high

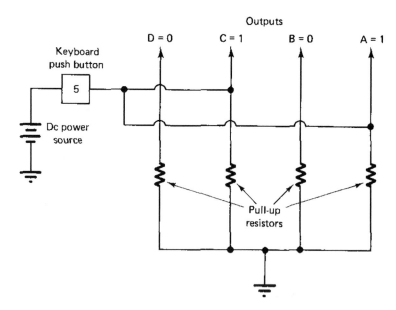

Figure 4-1. Simple input encoder

input and an active high output.

The encoder circuit of Figure 4-1 is very simple to construct and easy to understand. To make a 4-bit encoder of this type, simply connect a 1 to the desired output lines. When another input number is energized, this circuit has some problems. It cannot distinguish between different numbers when both inputs are activated. Figure 4-2 shows an encoder with two numbers applied to the input at the same time. Note that the output is still 0101 when the decimal number 4 is energized. This circuit does not effectively distinguish between two number values.

To make an encoder that will distinguish between different input numbers, we need to place diodes in the input lines. The diodes will be reverse biased to voltage values from the other inputs. As a result of this type of circuitry, the encoder can be used with multiple input lines. Its output will distinguish between each input and produce a representative output. This type of encoder is called a *diode matrix circuit*. It produces an output when a respective input number is actively high. This type of circuit would respond very well to the push buttons of a calculator keyboard.

Figure 4-3 shows an active-high diode matrix encoder. The decimal numbers of a calculator keyboard are represented by a push-button num-

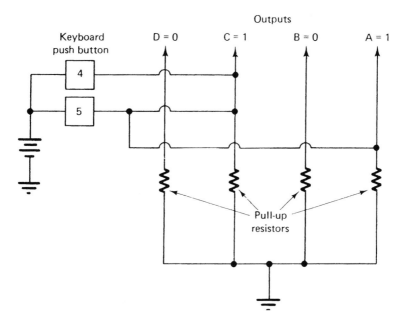

Figure 4-2. Two-input encoder

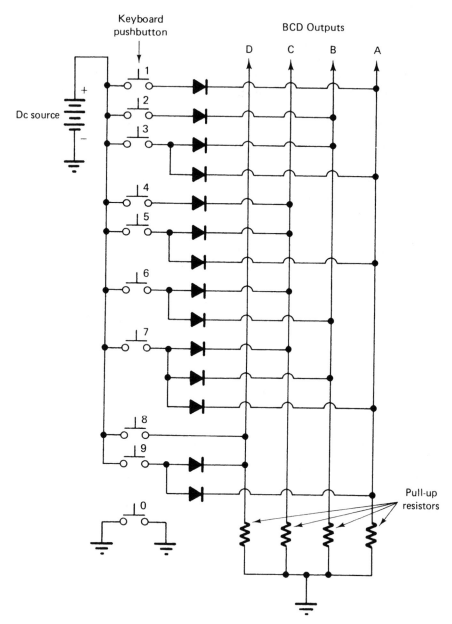

Figure 4-3. An active high diode matrix encoder

ber. The diodes of each input form an OR gate for the output. When a diode is forward biased, or 1, it produces a 1 output.

Another variation of the diode encoder matrix is shown in Figure 4-4. This encoder has an active low input and an active high output. The diodes attached to the input are reversed. This means that the cathode must be negative and the anode positive to produce conduction. When a push button is depressed, it supplies the necessary negative polarity to its diodes. The diodes conduct, causing the respective output lines to be negative, or 0. Pull-up resistors are attached to the positive side of the source. This means that a respective output line will be positive, or 1, unless it is forced to be low by diode conduction.

Assume now that the decimal 9 key is depressed on the active low diode matrix encoder of Figure 4-4. This action causes *D* and *A* output lines to be 1 and output lines *C* and *B* to be low. Notice that lines *C* and *B* are connected to ground,

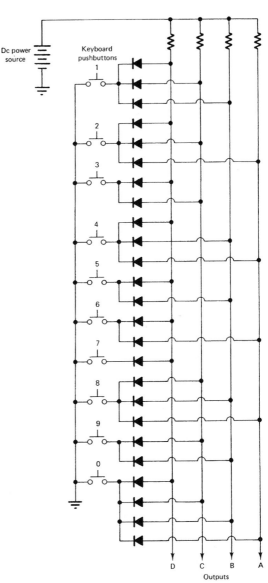

Figure 4-4. Active low input and active high output encoder

or the negative side, of the power source when this action is performed. Lines D and A are not connected to ground, which means that they will float high. In a sense the input lines conduct low data (active low) and the nonconductive diodes cause the output lines to go high. Diodes connected in this manner form the equivalent of an AND gate. Active low inputs and in some cases active low outputs are widely used in encoder ICs.

An OR gate encoder is shown in Figure 4-5. This encoder accomplishes the same function as the diode matrix active high input circuit of Figure 4-4. The respective OR gates are conductive when a 1, or high level, is applied to an input. The output will be high or low according to the conduction level of respective OR gates. No pull-up resistors are needed in this circuit configuration. Note that the OR gates are fairly complex. They have two or five inputs, according to their positions in the circuit. In some cases, a single number may cause only one input to be made active. Note the connections for inputs 1, 2, 4, and 8. The other numbers produce output that is determined by conduction of two or more gates. The 0 input line is not attached to any gate. When it is depressed, the output is a single 0. This type of encoder is considered to have an active high input and output. As a rule, multiple-input OR gates are somewhat expensive to accomplish in IC construction. Combination NOT, OR, and AND gates are widely used in the construction of IC encoders.

IC Encoders

Today encoders are formed on ICs. This type of encoder is generally quite complex compared with the diode matrix type of circuit. Their circuitry involves a number of combination logic gates. Typically NOT, AND and NOR gates are used in the combination logic circuits. The encoder is used according to its input and output connections. The user needs to be familiar with the pin connections and the general function of the device.

Figure 4-6 shows part of a data sheet for a 10-line decimal to 4-line BCD encoder. The 74147 9-input priority encoder accepts data from nine active low inputs. The inputs of this device are labeled $\overline{I}_1$ through $\overline{I}_9$. The overbar identifies the input as an active low. The output is an active low. Output is identified as A_3, A_2, A_1, and A_0. The output represents a BCD number with four lines. A priority is assigned to each input. When two or more inputs occur at the same time, the input with the highest priority appears in the output. The number $\overline{I}_9$ holds the highest priority. The priority diminishes with the values of the numbers.

The 74147 is described as a 10-line to 4-line priority encoder. It does

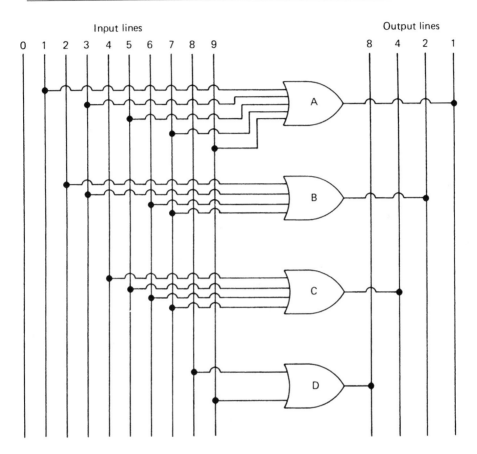

Figure 4-5. OR gate encoder

not have a zero input line. A 0 is encoded when all nine input data lines are high. This forces the four output lines to be high. Encoders of this type are primarily used to convert decimal keyboard inputs to a BCD output.

DIGITAL ELECTRONIC DISPLAYS

The average person looking at the display of a digital electronic device will not necessarily be familiar with binary numbers, the BCD method of display, octal, or hexadecimal codes. A digital system must therefore change its information into something that can be readily used without causing confusion. A function of nearly all digital systems is the changing

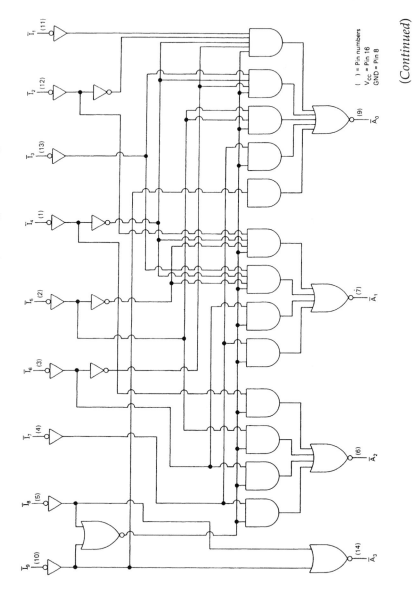

Figure 4-6. Encoder data sheet. (Courtesy of Signetics Corporation)

(Continued)

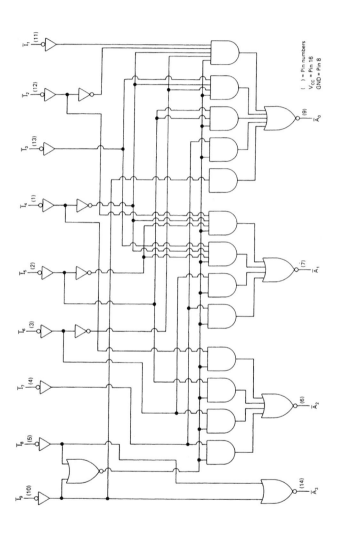

of signal information into decimal values that can be suitably displayed. The electronic display section of a digital system is responsible for this function.

The seven-segment method of display is used extensively in electrical systems and in it numerals are divided into seven segments, or slits. Figure 4-8 shows a display of this type. Illumination of two or more of these segments in an appropriate combination will produce the numbers 0 through 9. If, for example, segments *f, g, b,* and *c* are energized, the number 4 is displayed. Energizing all seven segments produces the number 8. Displays of this type are generally housed in a 14-pin dual-in-line IC package.

The illumination of a seven-segment display is achieved in a variety of ways. In a light-emitting diode, or LED, display, several discrete

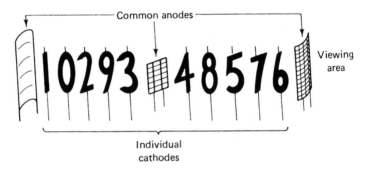

Figure 4-7. Exploded view of an individual numeral display

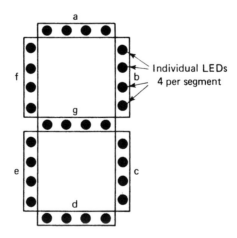

Figure 4-8. A seven-segment display with four diodes per segment

LEDs are commonly attached to a segment. In a common-anode type of display, the anodes of individual LEDs in the segment are connected and all segments are commonly connected together. The positive side of a 5-V source is applied to this connection. The common cathode of each segment is connected to the negative, or ground, of the power supply through a current-limiting resistor. A decoder IC such as an SN74LS47 is used to generate a digital signal that will illuminate the segments of the display. Figure 4-9 shows a diagram of the segments of an LED display. Note that the anodes of individual LEDs are connected together and tied to the + V_{CC} source. This type of display is illuminated when a logic 0 is applied to its cathode. A common-cathode version of the seven-segment display is also available for use. This type of display necessitates an active high, or logic 1, to each segment to cause it to be illuminated. A 74LS48 decoder IC is used to generate a digital signal that will illuminate the segments of the display.

Liquid crystal seven-segment displays are also used in many digital systems today. The configuration of this display is primarily the same as that of the LED unit. Illumination by the liquid crystal method is quite unique when compared with other displays. Most LCDs operate by the so-called twisted-nematic effect. These displays are powered through a special segment-addressing technique that reduces the number of connections to the display. The twisted-nematic type of LCD has a layer of liquid crystal sandwiched between two pieces of polarized glass and a set of electrodes with seven-segment columns, as shown in Figure 4-10. The polarized glass plates are designed to be 90° out of phase with each other. Normal light going through the top polarizer plate and into the liquid crystal undergoes a 90° twist as it passes through the material. This 90° twist permits light to pass through the polarizer electrode segments, liquid crystal, and to the bottom polarizer plate. However, applying an electric field to specific segments of the front and back electrodes will destroy the 90° twisting effect. The second polarizer then blocks light passing through the first polarizer. The end result is a display that appears as a dark area on a light background. The dark area is representative of the selected segments of the display. Using this operating principle, displays can be formed to produce letters, punctuation marks, and symbols, as well as numbers. As a rule, LCDs consume power in the microwatt range and are energized by low-voltage dc. A major disadvantage of the LCD is that it must operate in an area that has a reasonable level of ambient light to produce a suitable contrast in the display area.

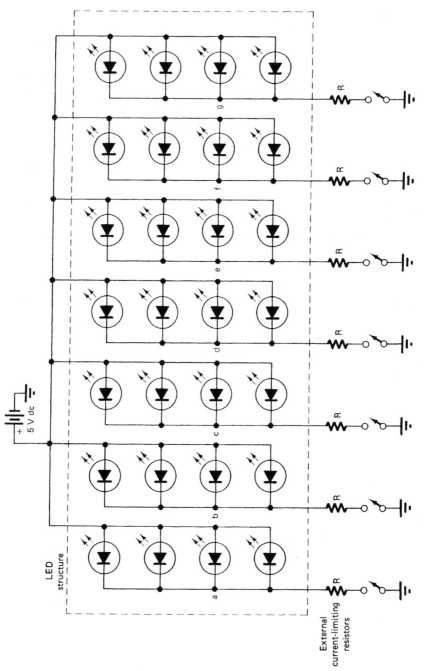

Figure 4-9. Segment structures of a seven-segment LED display

ALPHANUMERIC DISPLAYS

Letters, numbers, punctuation marks, and symbols for communication with a computer are frequently produced by alphanumeric displays. This display uses the 5 x 7 dot-matrix type of construction. Figure 4-11 shows the layout of this display.

The 5 x 7 dot matrix display uses 30 discrete LEDs in its construction. A specific LED is controlled by a combination of two switches. If the switches in row 4 and column 5 are both turned on at the same time,

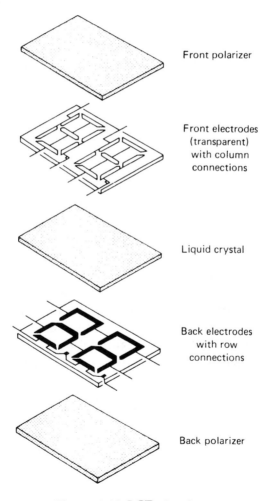

Front polarizer

Front electrodes
(transparent)
with column
connections

Liquid crystal

Back electrodes
with row
connections

Back polarizer

Figure 4-10. LCD structure

diode 25 is energized. If a complete vertical row is to be energized, it requires one column switch and all seven row switches to be actuated. A complete horizontal row is energized by one row switch and all five column switches. Through 12 different switching combinations, 96 characters can be produced by the alphanumeric display. These switching combinations are transmitted to the display unit through a 7-bit ASCII code. Recall that this code has a 3-bit group and a 4-bit group. The first 3 bits of data in the leftmost position of the 7-bit number are called the column-select group. The next 4 bits of data represent the row-select data. Figure 4-12 shows the ASCII code. Note that the column numbers are 000 through 111. The letter A, for example, is 100 0001. The letter *B* is 100 0010. An ASCII code is generated by pressing the letters and numbers of a typewriter keyboard. Integrated circuits accept the ASCII and translate it into a switching combination that will energize the appropriate segments of the dot matrix.

DECODING

Before a display device can be effectively used to develop a digital number, it must receive an appropriate signal from the counter circuit. The counter signal usually contains information in binary form. This information must be decoded so that it will energize the display device when a specific number occurs. Decoding is achieved by a number of four input gates connected to the A, B, C and D outputs of a BCD counter. When an appropriate binary number signal appears at the input of the decoder, it energizes the display device. In an actual circuit, a decoder connects the common or ground of the power source to seven specific bar segments to produce a display. In a sense, the device completes the manual switching operation of Figure 4-9 electronically. The decoder completes this operation automatically when it receives a suitable signal.

Two distinct types of decoders are available today for driving display devices. The discrete number display requires a decoder that has 10 distinct output signal possibilities, or a decimal output. In this type of decoding operation, only one output is energized at a time. This decoder generally has a BCD, or binary, input and a 10-line output. It is often called a 4-line to 10-line decoder. The seven-segment type of decoder is uniquely different. It produces two or more output signals when it ener-

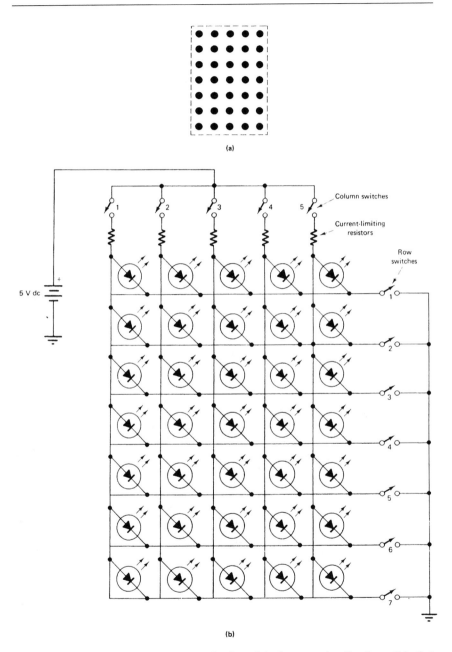

(a)

(b)

Figure 4-11. 5 x 7 LED display device: (a) dot-matrix display; (b) dot-matrix diagram

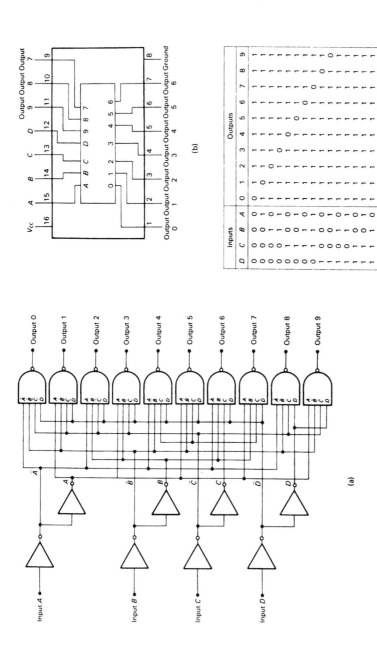

Figure 4-12. BCD to decimal decoder IC: (a) logic and connection diagrams; (b) dual-in-line and flat package; (c) truth table. (Courtesy National Semiconductor Corporation.)

gizes the display. When the number eight is displayed, all seven outputs must be energized to actuate the display. The basic method of actuating specific gates in the decoder is very similar for both decoder types.

BCD to Seven-Segment Decoders

The logic block of a BCD to seven-segment decoder is shown in Figure 4-13. The entire circuit of this functional block is built on a single IC chip numbered 74LS47. Inputs A, B, C, and D are applied to either one or two inverters as in the decimal decoder. The decoding process is very similar in nearly all respects to that of the decimal decoder. The seven-segment decoder, however, necessitates two or more outputs generated at the same time according to the number being displayed. In this decoder each NAND gate output produces a 0, or ground, when it receives four 1s at the input. This must be done for each of the respective seven-segment number combinations. A circle attached to each output indicates that it is active low. The output of this decoder is used to drive a common-anode seven-segment display.

The different decoding combinations needed to display the decimal numbers of a seven-segment display are shown in the truth table of Figure 4-13c. The A, B, C, and D inputs and corresponding a, b, c, d, e, f, and g outputs reflect the possible decoding combinations needed to produce a specific number output. There are three pins on the 74LS47 that we have not identified. Pin 3 is the lamp test (LT). When it goes to an active low (0), all segments illuminate regardless of the input. The BI/RBO (pin 4) input is an active low connection. When this input goes low, all segments are extinguished. This causes the display to go blank. The blanked condition can be tied to the next significant number of the display. The R_{BI} or R_{Bin} input (pin 5) of the 74LS47 serves as the blanking input of the display. As a rule, these pins are interconnected to other displays for simultaneously blanking of the entire display. The output of this decoder is used to actuate a common-anode LED display. As a rule, decoders with a seven-segment display are more commonly used in digital electronic systems than is the decimal decoder-type unit.

MULTIPLEXERS

A multiplexer acts like an electronically controlled rotary switch. A rotary switch is a mechanical device with a wiper arm that brushes across

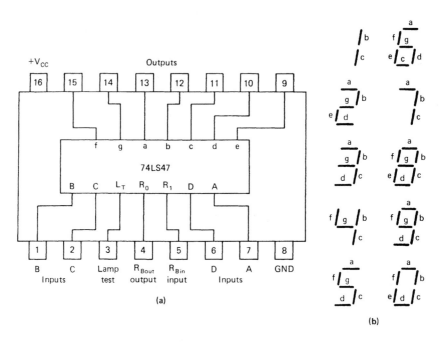

(a)

(b)

BCD inputs				7-segment outputs						
A	B	D	D	a	b	c	d	e	f	g
0	0	0	0	0	0	0	0	0	0	1
0	0	0	1	1	0	0	1	1	1	1
0	0	1	0	0	0	1	0	0	1	0
0	0	1	1	0	0	0	0	1	1	0
0	1	0	0	1	0	0	1	1	0	0
0	1	0	1	0	1	0	0	1	0	0
0	1	1	0	0	1	0	0	0	0	0
0	1	1	1	0	0	0	1	1	1	1
1	0	0	0	0	0	0	0	0	0	0
1	0	0	1	0	0	0	0	1	0	0

0 output = grounded element
1 output = open circuit

(c)

Figure 4-13. BCD to seven-segment decoding: (a) IC connection diagrams; (b) numbers and corresponding segments; (c) truth table.

several stationary contacts, connecting with each one as it changes position. Figure 4-14(a) shows the symbolic representation of a rotary switch. In this switch circuit, the wiper arm selects which one of the sources it will receive. The position of the wiper arm is changed manually. A multiplexer responds in the same manner. The switching operation is achieved electronically instead of manually. A select input is used to determine which of the data-input lines is to be transferred to the output. Figure 4-14(b) shows a multiplexer representation of the rotary switch. The switching operation of a multiplexer is achieved internally with logic gates. It has no moving parts and the transfer of data is achieved automatically.

A multiplexer is responsible for the distribution of data from one point to another in the operation of a digital system. The multiplexer, which is abbreviated MPX or MUX, is widely used as a data selector. It can be as simple as the two-input multiplexer of Figure 4-15. For this circuit, the data inputs are I_0 and I_1 and the select input is S. The logic level applied to S (which can be either 1 or 0) determines which AND gate is selected for transfer to the OR gate for output. The Boolean expression for this circuit is $Z = I_0S^* + I_1S$. When S is 0, $Z = I_0$, and when S is 1, the output $Z = I_1$. This shows that the select line controls which input is steered to the output.

MPXs are available today that have 4, 8, and 16 data-input lines. The number of data-selector bits required to produce control is a function of the data-input line number. A 16-line to 1-line multiplexer has a 4-bit data selector. The 4-bit selector is capable of producing a count of 0000 to 1111_2, or 0 to 15_{10}. The input has 16 lines numbered 0 through 15_{10}. The input line that is selected for transfer to the output is controlled by a 4-bit data selector. Operation is achieved by supplying a 4-bit number to the data selector and seeing the data at the selected input transferred to the output automatically.

74LS151 Multiplexer

A basic multiplexer is shown in Figure 4-16. The 74LS151 is an eight-line to one-line multiplexer. The logic diagram of this device shows the inputs numbered I_0, through 1_7. The data select lines are S_0, S_1, and S_2. The output is Y or Y^*. The enable input (E^*) determines the response of the output. When E^* is high, Y^* is high and Y is low. When E^* is low, the Y output is determined by the selected input line. Some manufacturers call the enable input a strobe input. Strobe generally refers to a fast, low-duty-cycle pulse that enables the device. The terms strobe and enable can be used

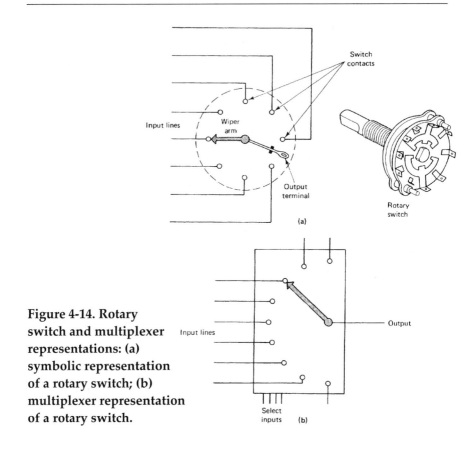

Figure 4-14. Rotary switch and multiplexer representations: (a) symbolic representation of a rotary switch; (b) multiplexer representation of a rotary switch.

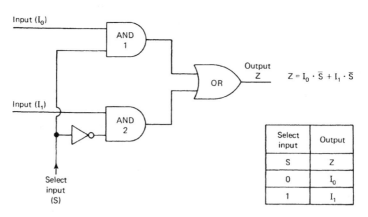

$$Z = I_0 \cdot \overline{S} + I_1 \cdot \overline{S}$$

Select input	Output
S	Z
0	I_0
1	I_1

Figure 4-15. Two-input multiplexer

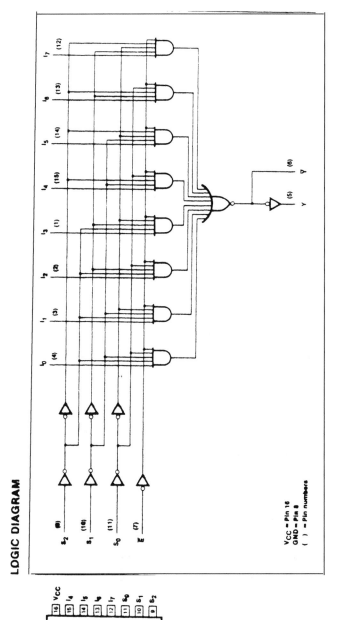

Figure 4-16. Data sheet of a 74LS151 multiplexer

(Continued)

FUNCTION TABLES

INPUTS												OUTPUTS	
E	S₂	S₁	S₀	I₀	I₁	I₂	I₃	I₄	I₅	I₆	I₇	Ȳ	Y
H	X	X	X	X	X	X	X	X	X	X	X	H	L
L	L	L	L	L	X	X	X	X	X	X	X	H	L
L	L	L	L	H	X	X	X	X	X	X	X	L	H
L	L	L	H	X	L	X	X	X	X	X	X	H	L
L	L	L	H	X	H	X	X	X	X	X	X	L	H
L	L	H	L	X	X	L	X	X	X	X	X	H	L
L	L	H	L	X	X	H	X	X	X	X	X	L	H
L	L	H	H	X	X	X	L	X	X	X	X	H	L
L	L	H	H	X	X	X	H	X	X	X	X	L	H
L	H	L	L	X	X	X	X	L	X	X	X	H	L
L	H	L	L	X	X	X	X	H	X	X	X	L	H
L	H	L	H	X	X	X	X	X	L	X	X	H	L
L	H	L	H	X	X	X	X	X	H	X	X	L	H
L	H	H	L	X	X	X	X	X	X	L	X	H	L
L	H	H	L	X	X	X	X	X	X	H	X	L	H
L	H	H	H	X	X	X	X	X	X	X	L	H	L
L	H	H	H	X	X	X	X	X	X	X	H	L	H

H = HIGH voltage level
L = LOW voltage level
X = Don't care

Figure 4-16 (*Continued*). Data sheet of a 74LS151 multiplexer interchangeably. It is important to note that the enable input of this device is an active low. The function table shows the first step having E^* high. This causes the Y output to be low regardless of the input data. When E^* is low, the Y output is determined by the data at the selected input. This is shown by the remaining steps of the function table. Operation is a process of supplying data to the appropriate input, selecting the desired input to be used, and transferring data to the output when the appropriate enable level is applied.

MULTIPLEXER APPLICATIONS

Multiplexers are important in the operation of a digital system. Applications include such things as data selection, data routing, parallel-to-series conversion, serial word generation, and logic function generation. We now take a look at some of these applications.

Logic Function Generation

Multiplexers are often used to implement Boolean expressions directly from a truth table without simplification. In this application the select inputs are identified as the logic variables. The data at each input are either high (1) or low (0) to satisfy the truth table.

Figure 4-17 shows how an eight-line multiplexer is used to implement a logic circuit that satisfies the truth table of a given Boolean expression. The input variables of the expression are attached to the select inputs S_0, S_1, and S_2, respectively. The level of these inputs determine which data input will appear at output Y. The input data are fixed at 1 or 0 according to the needs of the Boolean expression. Note that the input data lines are permanently attached to the $+V_{CC}$ source for a logic 1 and to ground for a 0 level.

Suppose now that we would like to implement the Boolean expression $Y = A\bar{B}C + A\bar{B}\bar{C} + ABC + \bar{A}BC$ with a multiplexer. A truth table for the expression is shown in Figure 4-17. This is a sum of products expression. The product of $A\bar{B}C$ is shown in the truth table. Note that a 1 appears only at step 4. The product of $A\bar{B}\bar{C}$ appears in the next column. A high or 1 appears only at step 1. A 1 is found only in step 7 for the product of ABC. For ABC, a 1 appears at step 6. To implement this expression, we must connect I_1, I_4, I_6, and I_7 to the positive side of the power source. Inputs I_0, I_2, I_3, and I_5 are connected to ground, or the low side, of the power source. When the select input switches from 100, 001, 110, and 111, the output at

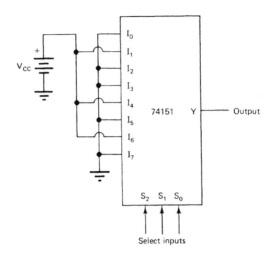

Truth table

Steps	C	B	A	$\overline{A}$	$\overline{B}$	$\overline{C}$	$(\overline{A} \cdot \overline{B} \cdot C)$	$(A \cdot \overline{B} \cdot \overline{C})$	ABC	$\overline{A}BC$	Y
0	0	0	0	1	1	1	0	0	0	0	0
1	0	0	1	0	1	1	0	1	0	0	1
2	0	1	0	1	0	1	0	0	0	0	0
3	0	1	1	0	0	1	0	0	0	0	0
4	1	0	0	1	1	0	1	0	0	0	1
5	1	0	1	0	1	0	0	0	0	0	0
6	1	1	0	1	0	0	0	0	0	1	1
7	1	1	1	0	0	0	0	0	1	0	1

Figure 4-17. Multiplexer logic function generator

Y will be $\overline{A}\overline{B}C + A\overline{B}\overline{C} + ABC + \overline{A}BC$. The select input determines the three variable input values for this Boolean expression. It should be noted that this method of implementation is certainly more efficient and easier to accomplish than using separate logic gates. This multiplexer circuit is good only for a three-variable expression. If a four-variable expression is to be implemented, the multiplexer would be changed to a 16-line device. The 74150 is a 16-input multiplexer.

Demultiplexers

Demultiplexers (DEMUX) perform the reverse of the multiplexing operation. A demultiplexer takes the data of a single input line and transfers them to one of many output lines. Demultiplexing is illustrated in Figure 4-18 as a rotary switch. The wiper arm of the switch is connected to the input and serves as the data-input line. The contacts of the switch are connected to the output lines. Operation is achieved by manually rotating the wiper arm to any one of the output contacts. The rotary switch demultiplexer operates manually and has moving parts and contact points. An IC de-multiplexer has no moving parts, and the output line is selected by logic gates. Data are transferred from the input data line to the output line by a binary code applied to the SELECT input. Demultiplexers are commonly used to distribute data from one circuit to another.

The 74LS138 Demultiplexer

Figure 4-19 shows a logic diagram of a one-line to eight-line 74LS138 demultiplexer. A single data-input line coming from the three-input enable gate is connected to each of the eight four-input AND gates. Two of the enable inputs $(E_1$ and $E_2)$ are active low and E_3 operates at an active high

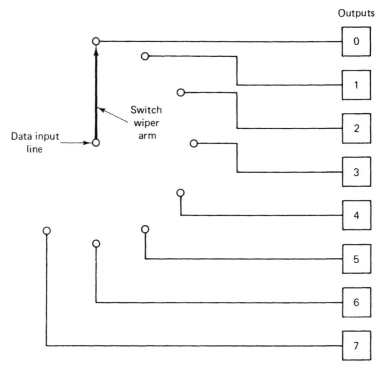

Figure 4-18. A rotary switch demultiplexer

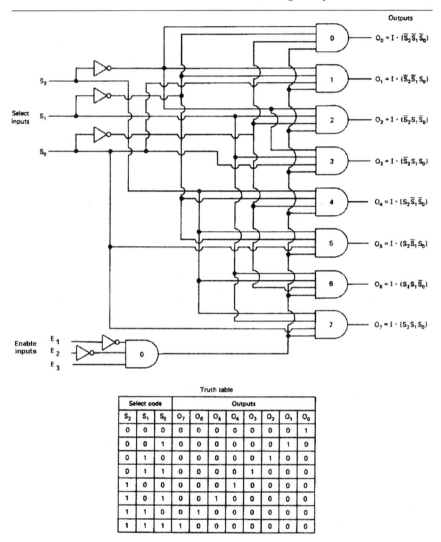

Figure 4-19. One-line to eight-line demultiplexer

level. The SELECT inputs S_0, S_1, and S_2 are attached to the remaining inputs of each AND gate. When a SELECT code is applied to S_0, S_1, and S_2, a specific output line is enabled and input (E_3) will appear at the selected output. If the SELECT input code is 000_2 for S_0, S_1, and S_2, AND gate 0 will be enabled, and output 0 will be a 1. Note that the SELECT input lines can be either a 1 or a 0 depending on the selected input code. A 000 input goes through three inverters before it is applied to the input of AND gate 0. The inversion process makes the three SELECT inputs all 1. If the data-input line is 1, the 0 AND gate is actuated and a 1 appears at output 0. A 0, or low, data input supplied to E_3 causes AND gate 0 to be inactive. The SELECT input code causes a specific output line to be actuated, which directs the input data to the selected output line. The accompanying truth table of Figure 4-19 summarizes the operation of the one-line to eight-line demultiplexer.

The one-line to eight-line demultiplexer of Figure 4-19 is very similar to the BCD to decimal decoder of Figure 4-12. The demultiplexer of Figure 4-19 has three SELECT inputs and one data-input line. If these inputs were combined, they could serve as the 4-bit BCD input of a decoder. An output of 0 to 7 could be derived from the input-select code. In many respects, a multiplexer is the same as a decoder. Some manufacturers may call this type of device a decoder/demultiplexer. It can be used for either function quite effectively. The 74LS138 is described as a decoder/demultiplexer.

DIGITAL COMPARATORS

An arithmetic operation that looks at two or more numbers and determines the relationship of these quantities is called a *digital comparator*. In its simplest form a comparator determines if two numbers are equal or if one value is greater than the other. The outputs of a comparator indicate if input A is greater than (>) input B, if input A is less than (<) input B, or if inputs A and B are equal (=). This arithmetic operation is different from that of addition or subtraction. A comparator looks at the magnitude of two input bits at the same time. The process starts with the MSB and works to the right, or to the least LSB. When one group has a binary 1 in one column and tile other does not, then the group with the 1 is larger. If all bits of the two number groups are the same in each position, then the two values are equal. Comparators are used to determine the magnitude of any binary code. In this presentation we are concerned only with applications that use natural binary numbers that have a positive value.

The exclusive-OR gate is a basic comparator that produces a 1 output when its two input bits are not equal. Because of this, the exclusive-OR gate has an inequality status. When two inputs are the same, the output will be 0. This gate can be used to compare two one-bit numbers in any place location. The output does not indicate which value is larger or smaller. A more sophisticated comparator tells the status of the output and

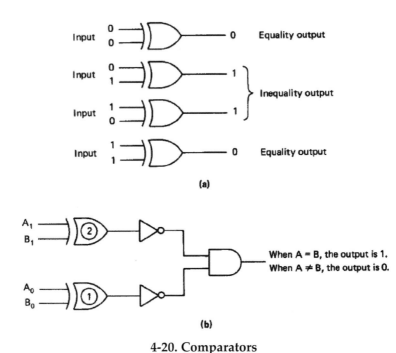

4-20. Comparators

which number group has the greatest value. Figure 4-20(a), shows the operations of a 2-bit exclusive-OR comparator.

In order to compare numbers that contain two or more bits, additional exclusive-OR gates are needed. Figure 4-20 shows a comparator diagram for two 2-bit numbers. In this diagram the two 2 LSBs are compared by exclusive-OR gate 1. These inputs are labeled A_0 and B_0. The next MSB is compared by exclusive-OR gate 2. These inputs are labeled A_1 and B_1. If the two-bit inputs are equal, the respective exclusive-OR gates will each produce a low, or 0, output. A 0 applied to the inverter will produce a 1 output. Two is applied to an AND gate produce a 1 output. This shows that when $A = B$, the output will be high, or 1. If either of the two-bit number sets are not equal, a 1 appears at the corresponding exclusive-OR gate output. This is inverted, causing a 0 to be applied to the input of the AND gate. If the inputs are not both 1, the output is 0. This shows when input A does not equal input B, the gate circuit registers a 0 output. As a result, this logic gate indicates a 1 for equality and a 0 for inequality of the 2-bit input numbers.

The basic 2-bit comparator of Figure 4-20 can be expanded to accommodate any number of bits. An exclusive-OR gate, inverter, and AND gate are needed for each 2-bit number set. The AND gate must accommodate all exclusive-OR gate outputs. A 4-bit number set necessitates a four-input AND gate. The AND gate determines the status of the circuit by comparing all of the inputs processed through the respective exclusive-OR inverter lines.

Exclusive-NOR Gates

The inversion function associated with the operation of an exclusive-OR gate in a comparator circuit has led to the development of an exclusive-NOR gate. The X-NOR is considered to be an equality comparator. Essentially, this logic gate inverts the output of an exclusive-OR gate. This means that when the inputs are equal, the output is 1. Unequal input values produce a 0 output. Figure 4-21 shows a logic symbol, a truth table, and an equivalent logic circuit for the exclusive-NOR gate. The output of an X-NOR gate is the complement of an X-OR gate. X-NOR gates are used primarily in arithmetic circuits. X-NOR gates are formed

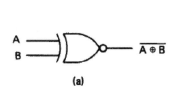

(a)

Input		Output	
B	A	X-OR	X-NOR
0	0	0	1
0	1	1	0
1	0	1	0
1	1	0	1

(b)

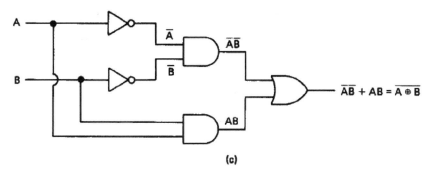

(c)

4-21. Exclusive-NOR gate

on ICs and are readily available today. As a rule, the functional use of an X-NOR gate is somewhat limited compared with the more popular X-OR gate. X-NOR gates are more widely used in comparator applications.

The Boolean expression for an exclusive-NOR can be derived from a truth table. The truth table of Figure 4-21(b) shows the output of an exclusive-NOR gate to be 1s when the inputs are equal. Transposing these outputs into a Boolean expression shows $\bar{A}\bar{B} + AB$ = X-NOR. The logic gate equivalent for this expression is shown in Figure 4-21(c). This gate circuit can be readily duplicated and formed on an IC. A 74LS266 is a quad two-input exclusive-NOR gate. Figure 4-22 shows the pin configuration of this chip. Exclusive-NOR gates are widely used in comparator circuit applications. You should be familiar with the functional response of this chip when it is encountered in a logic circuit. Its operation is often duplicated on other ICs to achieve this function.

Magnitude Comparators

Magnitude comparators are readily available in integrated circuit form. These comparators have outputs that show equality, greater-than, and less-than functions. The equality function for a one-bit comparison can be achieved with an exclusive-NOR gate. The Boolean expression of an X-NOR gate is $\overline{A_0 \oplus B_0}$. For a 2-bit comparator the expression is $A_1 A_0 = B_1 B_0$. This is expressed as $(\overline{A_0 \oplus B_0})\,(\overline{A_0 \oplus B_0})$. This is accomplished with two X-NOR gates and one AND gate. Figure 4-23 shows a 2-bit X-NOR

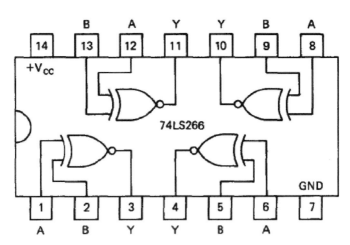

4-22. Pin connections of a 74LS266 X-NOR gate

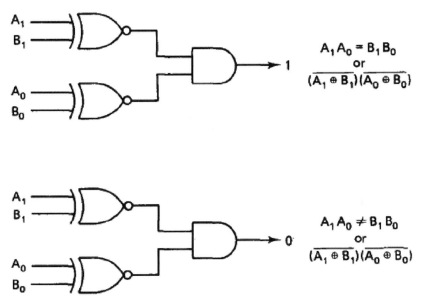

$A_1 A_0 = B_1 B_0$

or

$(\overline{A_1 \oplus B_1})(\overline{A_0 \oplus B_0})$

$A_1 A_0 \neq B_1 B_0$

or

$\overline{(A_1 \oplus B_1)(A_0 \oplus B_0)}$

4-23. Two-bit exclusive-NOR gate comparators

gate equality comparator. This circuit produces a high-level, or 1, output to indicate when the inputs are equal. Unequal inputs produce a low or 0, output.

Greater-than Function

The greater-than magnitude comparator function is somewhat difficult to accomplish. For a two-bit number the expression is $A_1 A_0 > B_1 B_0$. The first step in defining this function is an inspection operation. The MSBs A_1 and B_1 are inspected. If A_1 is 1 and B_1 is 0, then $A_1 A_0$ is greater than $B_1 B_0$ and the statement is true. A 1 is generated by the output. The operation is terminated at this point. If $A_1 = 0$ and $B1 = 1$, then the statement $A_1 A_0 > B_1 B_0$ is false. A zero appears in the output. The operation is again terminated at this point. However, if $A_1 = 1$ and $B_1 = 1$ or $A_1 = 0$ and $B_1 = 0$, the process moves to the next operational step. This step inspects the two LSBs. If $A_0 = 1$ and $B_0 = 0$, then the statement $A_1 A_0 > B_1 B_0$ is true. If $A_0 = 0$ and $B_0 = 1$, then the statement is false. The LSB decides the status of the output if the MSBs are equal. A combination of steps 1 and 2 decides the status of the output for $A_1 A_0 > B_1 B_0$.

A Boolean expression for the greater-than function of a magnitude comparator must be defined. The expression is derived from the inspection procedure previously outlined. $A_1A_0 > B_1B_0$ is expressed as $(A_1 > B_1)$ + $(A_1 = B_1)(A_0 > B_0)$. The first part of the expression $(A_1 > B_1)$ refers to the MSB inspection operation. The second part of the expression $(A_1 = B_1)$ refers to the equality function of the MSB. The third part $(A_0 > Bo)$ deals with the LSB. It shows the status of the LSBs.

The MSB inspection function $(A_1 > B_1)$ of our magnitude comparator can be achieved by an AND gate. The B_1 input is inverted first and applied to the input of the AND gate. $A_1\bar{B}_1$ shows when $A_1 > B_1$. The equality operation $(A_1 = B_1)$ is achieved by an X-NOR gate. $(\overline{A_1 \oplus B_1})$ shows when $A_1 = B_1$. The LSBs are then applied to an AND gate for the inspection process. The B_0 input is inverted and then applied to the input of an AND gate. AA, shows when $A_0 > B_0$. The outputs of the three respective logic circuits are then combined in an AND/OR gate combination. Figure 4-24 shows the structure of a 2-bit magnitude comparator that realizes $A_1A_0 > B_1B_0$.

Less-than Function

The logic statement of $A_1A_0 < B_1B_0$ is an expression of the less-than function of a magnitude comparator. This operation can be achieved with a circuit that is similar to the greater-than function. The first step in defining this operation is to identify the Boolean expression. Essentially, the function is an inspection operation. The MSBs are inspected first. If $A_1 = 0$ and $B_1 = 1$, then the statement A_1A_0 is less than B_1B_0 is true. A high (1) is generated at the < output. Operation is terminated at this time if this condition occurs. If $A_1 = 1$ and $B_1 = 0$, then the statement $A_1A_0 < B_1B_0$ is false. A low (0) appears at the < output. This again terminates the operation at this point. However, if $A_1 = 1$ and $B_1 = 1$ or $A_1 = 0$ and $B_1 = 0$, the inspec-

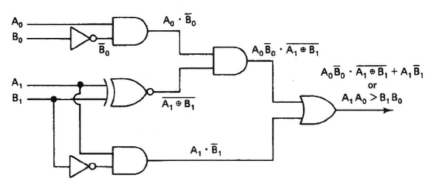

4-24. Two-bit comparator realizing A1A0 > B1B0

tion process moves to the next operational step. This step inspects the two LSBs. If $A_0 = 0$ and $B_0 = 1$, then the statement $A_1A_0 < B_1B_0$ is true. A 1 is generated at the < output. If $A_0 = 1$ and $B_0 = 0$, then the statement $A_1A_0 < B_1B_0$ is false. A 0 is generated at the < output. The LSB decides the status of the output if the MSBs are equal. A combination of steps 1 and 2 will decide the status of the output for $A_1A_0 < B_1B_0$.

A Boolean expression for the less-than function of a magnitude comparator can be defined from the previous inspection procedure. The statement $A_1A_0 < B_1B_0$ is expressed as $(A_1 < B_1) + (A_1 = B_1)(A_0 < B_0)$. The first part of the statement deals with MSB inspection. The second part deals with the equality function of the MSBs. The third part refers to the LSBs of the statement. The combined statement compares the magnitude of the MSB and LSB data.

The Boolean expression $A_1 < B_1) + (A_1 = B_1)(A_0 < B_0)$ must be changed into a logic gate diagram that will achieve this operation. The $(A_1 < B_1)$ part of the expression can be achieved with an AND gate. The A_1 input must be inverted to show the difference in magnitude. This shows that $(\bar{A}_1 B_1)$ achieves the operation of $A_1 < B_1$. The equality operation $(A_1 = B_1)$ is achieved by an X-NOR gate. $(\overline{A_1 \oplus B_1})$ shows when $A_1 = B_1$. The LSB data are then applied to an AND gate for the inspection process. A_0 is inverted and then applied to the input of the AND gate. $\bar{A}_0 B_0$ shows when $A_0 < B_0$. The three outputs are then combined in an AND/OR gate combination. Figure 4-25 shows the structure of a 2-bit magnitude comparator that indicates when $A_1A_0 < B_1B_0$. Note how different parts of the statement are achieved by the logic diagram.

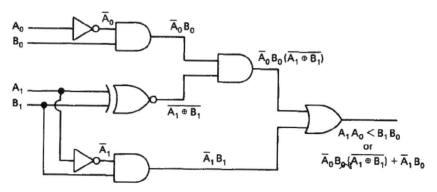

4-25. Two-bit comparator realizing A1A0 < B1B0

Four-bit Magnitude Comparator

The 74LS85 is a 4-bit magnitude comparator. It compares two 4-bit binary or BCD numbers and presents three possible magnitude indicators at its output. The 4-bit numbers are identified as A_3, A_2, A_1, A_0 and B_3, B_2, B_1, B_0. A_3 and B_3 are the MSBs, whereas A_0 and B_0 are the LSBs. A pin configuration and function table of the 74LS85 is shown in Figure 4-26.

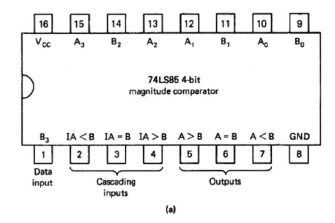

(a)

Comparing inputs				Cascading inputs			Outputs		
A_3, B_3	A_2, B_2	A_1, B_1	A_0, B_0	$I_{A>B}$	$I_{A<B}$	$I_{A=B}$	$A > B$	$A < B$	$A = B$
$A_3 > B_3$	X	X	X	X	X	X	H	L	L
$A_3 < B_3$	X	X	X	X	X	X	L	H	L
$A_3 = B_3$	$A_2 > B_2$	X	X	X	X	X	H	L	L
$A_3 = B_3$	$A_2 < B_2$	X	X	X	X	X	L	H	L
$A_3 = B_3$	$A_2 = B_2$	$A_1 > B_1$	X	X	X	X	H	L	L
$A_3 = B_3$	$A_2 = B_2$	$A_1 < B_1$	X	X	X	X	L	H	L
$A_3 = B_3$	$A_2 = B_2$	$A_1 = A_1$	$A_0 > B_0$	X	X	X	H	L	L
$A_3 = B_3$	$A_2 = B_2$	$A_1 = B_1$	$A_0 < B_0$	X	X	X	L	H	L
$A_3 = B_3$	$A_2 = B_2$	$A_1 = B_1$	$A_0 = B_0$	H	L	L	H	L	L
$A_3 = B_3$	$A_2 = B_2$	$A_1 = B_1$	$A_0 = B_0$	L	H	L	L	H	L
$A_3 = B_3$	$A_2 = B_2$	$A_1 = B_1$	$A_0 = B_0$	L	L	H	L	L	H
$A_3 = B_3$	$A_2 = B_2$	$A_1 = B_1$	$A_0 = B_0$	X	X	H	L	L	H
$A_3 = B_3$	$A_2 = B_2$	$A_1 = B_1$	$A_0 = B_0$	H	H	L	L	L	L
$A_3 = B_3$	$A_2 = B_2$	$A_1 = B_1$	$A_0 = B_0$	L	L	L	H	H	L

H = HIGH voltage level
L = LOW voltage level
X = Don't care

(b)

4-26. 74LS85 4-bit magnitude comparator: (a) pin configuration (b) function table

Operation of the 74LS85 is described by the function table of Figure 4-26. All the possible logic conditions are shown in the table. This device can be used as a single 4-bit comparator or be expanded to accommodate a larger number of bits. The upper part of the function table shows the normal operation under all conditions that will occur in a single device or in a series expansion application. In this part of the table the three outputs are mutually exclusive. The lower part of the table shows how the device will respond in a feed-forward parallel expansion application. The expansion inputs are identified as $IA > B$, $IA = B$, and $IA < B$. These serve as the LSB positions. When used for a series expansion application, the $A > B$, $A = B$, and $A < B$ outputs of the least-significant word are connected to the corresponding $IA > B$, $IA = B$, and $IA < B$ inputs of the next-higher stage. Four-bit comparators can be added in this manner to any length. As a rule, some delay is achieved by each device. The propagation delay of a 74LS85 is 15 ns for each 4-bit word.

SUMMARY

An important characteristic of all digital systems is communication. The system must be capable of communicating with an operator or another system. As a rule, the system uses data in the form of a code. It must accept an input signal and convert it to a usable code. This operational step is called encoding. Data are then distributed to different parts of the system. This is the transmission function. Data transmission involves the use of multiplexers and demultiplexers. Multiplexing is achieved with a device that can select one of a number of inputs and pass the logic level of that input on to the output. Demultiplexing incorporates a device that directs information from a single input to one of several outputs.

Encoders are used to convert one or more input data lines into multiline output data. A representative encoder converts decimal data into binary-coded-decimal data. The keyboard of a calculator is an example of a encoder. It employs a switch for each key that represents a distinct decimal value or math function. When a button or key is depressed, it produces a distinct decimal value for the system. This action may generate a code that has six or more bits of information.

An electronic display is responsible for converting digital signals into something that can be viewed by the operator. Seven-segment displays produce a number by illuminating seven segments or slits. Alphanumeric displays produce letters, numbers, punctuation marks, and symbols. This display uses a 5 x 7 dot matrix diode construction.

A multiplexer is responsible for the distribution of data from one point to another in the operation of a digital system. A multiplexer acts like an electronically controlled rotary switch. It makes connection between one or more inputs and the output. MPXs are available with 4, 8, or 16 data input lines with a 4-bit data-select line.

Demultiplexing, or DEMUX, is the reverse of multiplexing. Demultiplexing takes data from a single input line and transfers it to one or more output lines. DEMUX resembles a rotary switch that has the input connected to the wiper arm and the contacts serving as the output. An IC demultiplexer has no moving parts or contact points. Data are transferred to the output by selected logic gates.

Priority encoders are often used in keyboard applications. When a key is depressed, it energizes the input of an encoder. A single digit input such as the decimal number 5 responds as a switch. This switch actuates the input so that it produces a coded representation of the input. An encoder produces a coded version of the decimal input. This could be a BCD output. Essentially, the encoder changes a decimal value to a BCD value.

The 74147IC is a 10-line decimal to 4-line BCD priority encoder. It accepts data from nine active-low inputs and produces an active-low BCD output. This IC has nine active low inputs labeled 7_1 through L. The overbar indicates the active low status of the input. The BCD output is identified as A_2, A_2, A_1, and A0. These four lines indicate the active low status of the BCD output. The bar indicates the active low status of the output.

The term priority encoder is used in the description of the 74147. Priority refers to the value assignment of each input. When two or more inputs are simultaneously active, the input with the highest priority appears as the output. Input i_o has the highest priority and has the lowest. A priority input encoder has selection capabilities according to the weighted values of the input.

A digital comparator looks at two or more numbers and determines if the numbers are equal or if one value is greater than the other. An exclusive-OR gate can be used as a comparator. An X-OR gate is a inequality comparator. It develops a 1 output when the inputs are unequal. An exclusive-NOR gate is an equality comparator. It develops a 1 output when the inputs are equal. Magnitude comparators are available in ICs. These comparators show equality, greater-than, and less-than functions. An X-NOR gate is needed for each bit of data compared.

Chapter 5

Number Systems, Conversions and Codes

INTRODUCTION

Digital electronics is undoubtedly the fastest-growing area in the field of electronics today. Personal computers, calculators, watches, clocks, video games, test instruments, and home appliances are only a few of the applications. Many of these things were unheard of only a few years ago. Digital electronics now plays an important role in our daily lives.

A general discussion of electronics is primarily centered around analog applications. An analog device is one in which a quantity is represented on a continuous scale. Temperature, for example, can be determined by the position of a column of mercury. Voltage, current, and resistance can be determined by the movement of a coil of wire that interacts with a magnetic field. Analog devices are usually concerned with continuously changing values. An analog value could be any one of an infinite number of values. Radio, television, and stereo sound systems deal primarily with manipulation of analog data.

Digital electronics is considered to be a counting operation. A digital watch tells time by counting generated pulses. The resulting count is then displayed by numbers representing hours, minutes, and seconds. A computer also has an electronic clock that generates pulses. These pulses are counted and in many cases manipulated to perform a control function. Digital circuits can store signal data, retrieve it when needed, and make operational decisions. Signal values are generally represented by two-state data. A pulse is either present or it is not. Data are either of high value or low value, with nothing in between. Digital electronics is dependent on the manipulation of two-state data.

DIGITAL NUMBERING SYSTEMS

Digital information has been used by human beings during almost all of their history. Parts of the human body were first used as a means of counting. Fingers and toes were often used to represent numbers. In fact, the word *digitus* in Latin means finger or toe. This term is the basis of the word digital.

Most counting that we do today is based on groups of 10. This is probably an outgrowth of our dependence on fingers and toes as a counting tool. Counting with 10 as a base is called the *decimal* system. Ten unique symbols, or digits, are included in this system: 0, 1, 2, 3, 4, 5, 6, 7, 8, and 9. In general, the number of discrete values or symbols in a counting system is called the *base*, or *radix*. A decimal system has a base of 10.

Nearly all numbering systems have place value. This refers to the value that a digit has with respect to its location in the number. The largest number value that can be represented at a specific location is determined by the base of the system. In the decimal system, the first position to the left of the decimal point is called the units place. Any number from 0 to 9 can be used in this place. Number values greater than 9 are expressed by using two or more places. The next location to the left of the units place is the 10s position. Two-place numbers range from 10 through 99. Each succeeding place added to the left has a value that is 10 times as much as the preceding place. With three places, the place value of the third digit is 10 x 10 x 10, or 1000. For four places, the place value is 10 x 10 x 10 x 10, or 10,000. The values continue for 100,000, 1,000,000, 10,000,000, and so on.

Any number in standard form can be expressed in expanded form by adding each weighted place value. The decimal number 2319 is expressed as (1000 x 2) + (100 x 3) + (10 x 1) + (1 x 9). Note that the weight of each digit increased by 10 for each place to the left of the decimal point. In a number system, place values can also be expressed as a power of the base. For the decimal system, the place values are 10^0, 10^1, 10^2, 10^3, and so on. Each succeeding place has a value that is the next higher power of the base.

The base 10, or decimal, numbering system is extremely important and widely used today. Electronically, however, the decimal system is rather difficult to use. Each number would require a specific value to distinguish it from the others. Number detection would also require some unique method of distinguishing each value from the others. The electronic circuitry of a decimal system would be rather complex. In general, base 10 values are difficult to achieve and awkward to maintain.

BINARY NUMBERING SYSTEMS

Nearly all digital electronic systems are of the binary type. This type of system uses 2 as its base, or radix. Only the numbers 0 and 1 are used in a binary system. Electronically, only two situations, such as a value or no value, are needed to express binary numbers. The number 1 is usually associated with some voltage value greater than zero. Binary systems that use voltage as 1 and no voltage as 0 are described as having positive logic. Negative logic uses voltage for 0 and no voltage for 1. Positive logic is more readily used today. Only positive logic is used in this discussion.

The two operational states or conditions of an electronic circuit can be expressed as on and off. An off circuit usually has no voltage applied, so represents the 0, or off, state. An on circuit has voltage applied, so represents 1, or on, state. With the use of electronic devices it is possible to change states in a microsecond or less. Millions of is and Os can be manipulated by a digital system in a second.

A binary digit can be expressed as either 1 or 0. The term *bit* is commonly used to describe a binary digit.

The operational basis of a binary system is very similar to that of the decimal system. The base of the binary system is 2. This means that only the numbers 0 and 1 are used to denote specific numbers. The first place to the left of the binary point is the units, or is location. Place values to the left are expressed as powers of 2. Representative values are $2^0 = 1$, $2^1 = 2$, $2^2 = 4$, $2^3 = 8$, $2^4 = 16$, $2^5 = 32$, $2^6 = 64$, and so on.

When different numbering systems are used in a discussion, they usually incorporate a subscript number to identify the base of the numbering system being used. The number 110_2 is a typical expression of this type. The subscript number 2 is used to denote the base of the numbering system. The binary point, which follows the 0, is usually omitted. Thus 110_{10} is used to indicate when the number is expressed in decimal form. We will use other numbering systems, such as the base 8, or octal, system and the base 16, or hexadecimal, system. Numbers expressed in these bases will be identified with subscripts to avoid confusion with other numbers.

When only one numbering system is used in a discussion, the subscript notation is generally not needed.

A binary number such as 101_2 is the equivalent of the decimal number 5_{10}. Starting at the binary point, the digit values are $1 \times 2^0 + 0 \times 2^1 + 1 \times 2^2$ or $1 + 0 + 4 = 5$. The conversion of a binary number to a decimal value is shown in Figure 5-1.

A shortcut version of the binary-to-decimal conversion process is shown in Figure 5-2. In this conversion method, first write the binary number. In this example, 1001101_2 is used. Starting at the binary point, indicate the decimal value of each binary place containing a 1. Do not indicate a value for the zero places. Add the indicated place values. Record the decimal equivalent of the binary number. Practice this procedure on several different binary numbers. With a little practice the conversion process is very easy to achieve.

Changing a decimal number to a binary number is achieved by repetitive steps of division by the number 2. When the quotient is even with no remainder, a 0 is recorded. A remainder is also recorded. In this case, it will always be the number 1. The steps needed to convert a decimal number to a binary number are shown in Figure 5-3.

The conversion process is achieved by first recording the decimal value. The decimal number 30 is used in this example. Divide the recorded number by 2. In this case, 30/2 equals 15. The remainder is 0. Record the 0 as the first binary place value. Transfer 15 to position 2. Divide this

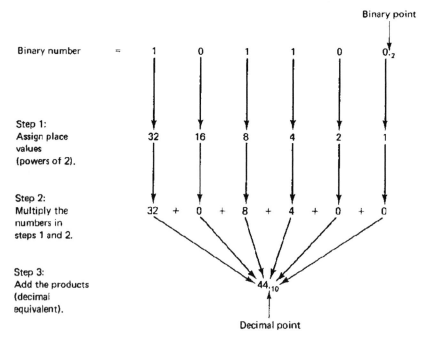

Figure 5-1. Conversion of a binary number to a decimal number

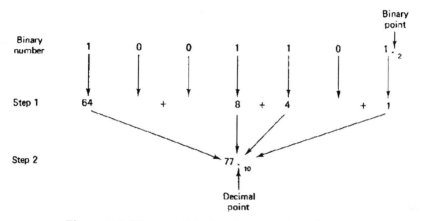

Figure 5-2. Binary-to-decimal conversion shortcut

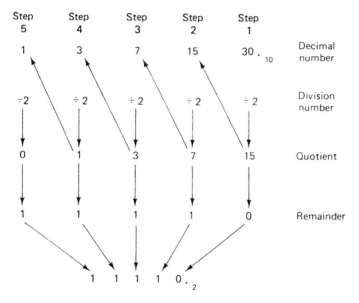

Figure 5-3. Conversion of a decimal equivalent number to a binary number

value by 2. 15/2 equals 7 with a remainder of 1. Record the remainder. Transfer 7 to position 3. 7/2 equals 3 with a remainder of 1. Record the remainder and transfer 3 to position 4. 3/2 equals 1 with a remainder of 1. Record the remainder and transfer 1 to position 5. 1/2 equals 0 with a remainder of 1. The conversion process is complete when the quotient has a value of 0. The binary equivalent of a decimal is indicated by the recorded remainders. For this example, the binary equivalent of 30_{10} is 111102.

Practice the decimal-to-binary conversion process with several different number values. With a little practice the process becomes relatively easy to accomplish.

Binary Fractions

Having discussed the whole numbers of a binary system, let us now direct attention toward fractional number values. The binary point of a number decides the difference between integer and fractional values. Binary whole numbers are located to the left of the binary point. Fractional values are located to the right of the binary point. Whole numbers are identified as positive powers while fractional numbers are expressed as negative powers.

A number such as 1101.101_2 has both integer and fractional values. The whole number to the left of the binary point is 1101. This represents 1 x 2^3 + 1 x 2^2 + 0 x 2^1 + 1 x 2^0 or $13_{,0}$. The fractional part of this number is positioned to the right of the binary point. It is $.101_2$. This represents the sum of negative exponent values and becomes 1 x 2^{-1} + 0 x 2^{-2} + 1 x 2^{-3} or 1/2 + 0/4 + 1/8. The decimal equivalent is 0.5 + 0.0 + 0.125 or $0.625_{,0}$. The number 1101.101_2 equals 13.625_{10}. The fractional part of this number is an even value of the decimal equivalent. If the binary values do not produce an even combination of the fractional number the number will have infinite length. As a rule, only three or four fractional places are needed to define most binary numbers.

Binary Coded Decimal Numbers

When large numbers are to be indicated by binary numbers, they become somewhat awkward and difficult to use. For this reason, the binary-coded decimal method of counting was devised. In this type of system, four binary digits are used to represent each decimal digit. To illustrate this procedure, we have selected the decimal number 329_{10} to be converted to a binary-coded decimal (BCD) number. As a binary number, 32910 = 101,001,0012.

To apply the BCD conversion process, the base 10 number is first divided into discrete digits according to place values (see Figure 5-4). The number 329_{10} therefore equals the digits 3-2-9. Converting each digit to binary permits us to display this number as 0011- 0010-1001_{BCD}. Decimal numbers up to 999_{10} can be displayed and quickly interpreted by this process with only 12 binary numbers. The dash line between each group of digits is extremely important when displaying BCD numbers. It is used to denote that the number is in binary-coded decimal form.

The largest decimal number to be displayed by four binary digits of a BCD number is 9_{18} or 1001_2. This means that six counts of the binary number are not being used. These are 1010, 1011, 1100, 1101, 1110, and 1111. Because of this the octal, or base 8, and hexadecimal, or base 16, numbering systems were devised. Digital systems still process numbers in binary form but usually display them in BCD, octal, or hexadecimal values.

Given decimal number		$329._{10}$	
Step 1 Grouping of digits	(3)	(2)	(9)
Step 2 Conversion of each digit to binary group	(0011)	(0010)	(1001)
Step 3 Combine group values		0011 / 0010 / 1001 $._{BCD}$	

Figure 5-4. Converting a decimal number to a BCD number

OCTAL NUMBERING SYSTEM

Octal, or base 8, numbering systems are commonly used to process large numbers through digital systems. The octal system of numbers uses the same basic principles outlined with the decimal and binary systems.

The octal numbering system has a base of 8. The digits 0, 1, 2, 3, 4, 5, 6, and 7 are used. The place values starting at the left of the octal point are powers of 8: $8^0 = 1$, $8^1 = 8$, $8^2 = 64$, $8^3 = 512$, and $8^4 = 4096$, and so on.

The process of converting an octal number to a decimal number is the same as that used in the binary-to-decimal conversion process. In this method, however, the powers of 8 are employed instead of the powers of 2. To convert the number 265_8 to an equivalent decimal number, see the procedure outlined in Figure 5-5.

Converting an octal number to an equivalent binary number is very similar to the BCD conversion process discussed previously. The octal number is first divided into discrete digits according to place value. Each octal digit is then converted into an equivalent binary number using only three digits. The steps of this procedure are shown in Figure 5-6. You may want to practice this conversion process to gain proficiency in its use.

Converting a decimal number to an octal number is a process of repetitive division by the number 8. After the quotient has been determined, the remainder is brought down as the place value. When the quotient is even with no remainder, a zero is transferred to the place position. The procedure for converting the number 4098_{18} to its octal equiva-

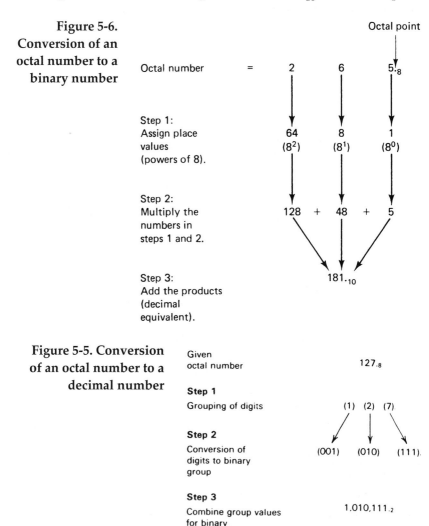

Figure 5-6. Conversion of an octal number to a binary number

Octal point

Octal number   =   2   6   5.$_8$

Step 1: Assign place values (powers of 8).

64   8   1
(8^2) (8^1) (8^0)

Step 2: Multiply the numbers in steps 1 and 2.

128 + 48 + 5

Step 3: Add the products (decimal equivalent).

181.$_{10}$

Figure 5-5. Conversion of an octal number to a decimal number

Given octal number

127.$_8$

Step 1
Grouping of digits

(1) (2) (7)

Step 2
Conversion of digits to binary group

(001) (010) (111)

Step 3
Combine group values for binary equivalent

1,010,111.$_2$

lent is outlined in Figure 5-7.

Converting a binary number to an octal number is a very important conversion process found in digital systems. Binary numbers are first processed through the equipment at a very high speed. An output circuit then accepts this signal and may convert it to an octal signal that can be displayed on a readout device.

Assume now that the number $10,110,101_2$ is to be changed into an equivalent octal number. The digits must first be divided into groups of three, starting at the octal point. Each binary group is then converted into an equivalent octal number. These numbers are then combined, while remaining in their same respective places, to represent the equivalent octal number. See the conversion steps outlined in Figure 5-8.

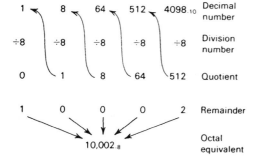

Figure 5-7. Conversion of a decimal number to an octal number

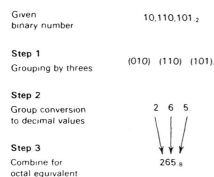

Figure 5-8. Conversion of a binary number to an octal number

HEXADECIMAL NUMBERING SYSTEMS

The hexadecimal numbering system is used to process large numbers. The base of this system is 16, which means that the largest value used in a place is 15. Digits used to display this system are the numbers 0 through 9 and the letters A, B, C, D, E, and F. The letters A through F are used to denote the digits 10 through 15, respectively. The place values of digits to the left of the hexadecimal point are the powers of 16: $16° = 1$, $16' = 16$, $16^2 = 256$, $16^3 = 4096$, $16^4 = 65,536$, $16^5 = 1,048,576$, and so on.

The process of changing a hexadecimal number to a decimal number is achieved by the same procedure outlined for other conversions. Initially, a hexadecimal number is recorded in proper digital order as outlined in Figure 5-9. The powers of the base are then positioned under the respective digits. In a hexadecimal conversion, step 2 is usually added to show the values of the letters. Each digit is then multiplied by its place value to indicate discrete place-value assignments. Steps 1 and 2 are then multiplied together. In step 3, these product values are added, giving the decimal equivalent of a hexadecimal number in step 4.

The process of changing a hexadecimal number to a binary equivalent is a simple grouping operation. Figure 5-10 shows the operational steps for making this conversion. Initially, the hexadecimal number is separated into discrete digits in step 1. Each digit is then converted to an equivalent binary number using only four digits per group. Step 3 shows the binary groups combined to form the equivalent binary number.

The conversion of a decimal number to a hexadecimal number is achieved by the repetitive division process used with other number systems. In this procedure, however, the division factor is 16 and the remainders can be as large as 15. Figure 5-11 shows the necessary procedural steps for achieving this conversion.

Converting a binary number to a hexadecimal equivalent is a reverse of the hexadecimal to binary process. Figure 5-12 shows the fundamental steps of this procedure. Initially, the binary number is divided into groups of four digits, starting at the hexadecimal point. Each grouped number is then converted to a hexadecimal value and combined to form the hexadecimal equivalent.

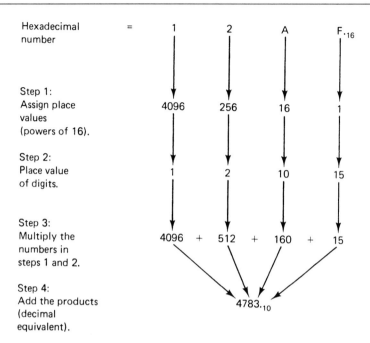

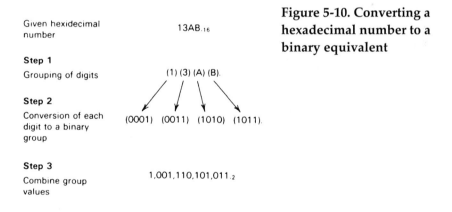

Figure 5-9. Conversion of hexadecimal number to a decimal number

Given hexidecimal number   13AB₁₆

Figure 5-10. Converting a hexadecimal number to a binary equivalent

Step 1

Grouping of digits   (1) (3) (A) (B).

Step 2

Conversion of each digit to a binary group   (0001) (0011) (1010) (1011).

Step 3

Combine group values   1,001,110,101,011.₂

BINARY CODES

Digital systems are binary operated and use is and 0s as data for signal processing. However, individuals communicate with each other in decimal numbers and alphabetic characters. When working with a digital system we need some method of interfacing between binary data and

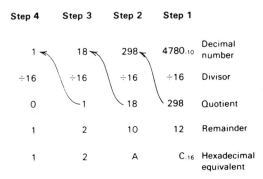

Step 4	Step 3	Step 2	Step 1	
1	18	298	4780.10	Decimal number
÷16	÷16	÷16	÷16	Divisor
0	1	18	298	Quotient
1	2	10	12	Remainder
1	2	A	C.16	Hexadecimal equivalent

Figure 5-11. Conversion of a decimal number to a hexadecimal number

alphanumeric data. To accomplish this a number of binary codes have been developed. The process of generating these codes is called *encoding*. The process of recognizing or converting a binary code to decimal or alphanumeric characters is called *decoding*.

In our discussion of digital systems up to this point, we have had some experience with codes. The numbering systems (binary, octal, and hexadecimal) are actually codes. They permit us an opportunity to change data of one form into another form for easy manipulation. There are other codes and data-transmission procedures used in digital systems today.

Decimal	Code
	Weight 8421
0	0000
1	0001
2	0010
3	0011
4	0100
5	0101
6	0110
7	0111
8	1000
9	1001

Figure 5-12. Decimal number to BDC values

BCD Code

The BCD code is very popular in digital systems. This coding procedure was discussed as a method of displaying binary data. The BCD code is considered to be a weighted code. Each bit of data carries weight according to its location in the numerical display. This progresses in natural binary form from 0000 to 1001_2. It is a common practice to describe BCD numbers as the 8421 code because of its weighted structure. Figure 5-12 shows a table of BCD numbers and their decimal equivalents.

Parity Code

The parity code is an error-checking code in which binary numbers are transmitted with an additional bit that is used only for error detection. In this code all data are transmitted either with an odd number of is *(odd parity)* or an even number of is *(even parity)*. The odd-parity code is more commonly used, since it is not possible to transmit a number or word with all Os. Figure 5-13 shows a display of data and the generation of a parity bit. Both odd and even parity codes are shown. Note that a parity bit is generated according to the number of ls in a number. In the odd-parity code, a 0 parity bit is generated when the number of is in a number is odd. A 1 parity bit is generated when the number of is even. The even-parity code is just the opposite of the odd code.

Odd parity		Even parity	
Data	Parity bit	Data	Parity bit
001001	1	001001	0
110100	0	1110110	1
1011011	0	101111	1
101010	0	100110	1
110011	1	100111	0

Figure 5-13. Odd parity and even parity code tables

The parity code is designed to warn the system that a particular binary number is in error. It does not detect an actual error. It is used to respond to the data input of a digital system. A parity code can be used with other techniques to detect an actual 1 or 0 error in a data display. This usually permits the system to display a flag or indicate an error *sign* when something occurs improperly. Parity codes are widely used to evaluate keyboard data entry into a system.

Gray Code

The Gray code is a method of displaying binary data with only a change of 1 bit of *data* between each number. It is used to encode data such as the shaft position of a motor or the physical location of a moving object. A Gray-code chart is shown in Figure 5-14. This code is compared with an equivalent binary number.

Figure 5-15 shows an eight-segment shaft encoder of the binary type and the Gray-code type. Assume that a light sensor is located at a specific position to detect positional changes. If the wheel position is such that the center of a segment is directly opposite to the light sensor, the output will read the position of the location correctly. Suppose, however, that the position of the shaft is as shown in Figure 5-15. This will indicate the position as either 000 or 111. It is possible for the sensor to be confused in this position and give an erroneous indication. It could be all 1s or all 0s with a very fine line of difference. This could produce an error that is the difference between maximum and minimum shaft position. This method of display is very confusing as a position indicator. Now consider the Gray-code encoder of Figure 5-15. Note that the two outer rings of the display have the same color in adjacent segments. This means that the only possible point of confusion is in the inner ring. The error possibility at this location is only one segment position. This represents a 1-bit change for each decimal value. An error of this type is not particularly significant. The Gray code display is more accurate and less prone to error indications.

A shaft encoder may have as many as 12 Gray-coded rings. Since 2^{12} = 4096, it is possible for this type of encoder to resolve angular positions within approximately 0.1°. Gray-code-position detection is widely used today in digital systems that control robots or computer controlled manufacturing machinery.

Alphanumeric Code

To communicate with a digital system or a computer, we need numbers, letters, and other symbols. This information must be changed into some type of standardized code. Codes that represent this type of information are called *alphanumeric*. They must represent 10 decimal digits, 26 letters of the alphabet, and several punctuation marks. One very important code is the American Standard Code for Information Interchange (ASCII). This is a 7-bit code. It has a 3-bit group and a 4-bit group of binary numbers. The 3-bit group is a column-select group and consists of the most significant or first 3 bits on the left. The four-bit group is the row-select binary number. Figure 5-16 shows the ASCII code. Note that the column numbers are 000 through 111. The letter A, for example, is 100 0001. This shows its location as column 1Q0 and row 0001. The number 4 is 011 0100. When an ASCII keyboard is used, a 7-bit code occurs for each letter or code button depressed. Note also that there are several abbreviations included in the ASCII code. These are defined in the chart.

Decimal number	Gray code	Binary number
0	0000	0000
1	0001	0001
2	0011	0010
3	0010	0011
4	0110	0100
5	0111	0101
6	0101	0110
7	0100	0111
8	1100	1000
9	1101	1001
10	1111	1010
11	1110	1011
12	1010	1100
13	1011	1101
14	1001	1110
15	1000	1111

Figure 5-14. Gray code table with binary and decimal numbers

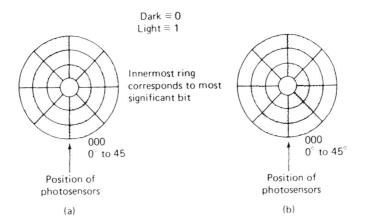

Dark ≡ 0
Light ≡ 1

Innermost ring corresponds to most significant bit

000
0° to 45

Position of photosensors

(a)

000
0° to 45°

Position of photosensors

(b)

Figure 5-15. Eight-segment shaft encoder: (a) binary coded wheel; (b) gray code wheel.

American Standard Code for Information Interchange.

	000	001	010	011	100	101	110	111
0000	NULL	DC_0	b	0	@	P		
0001	SOM	DC_1	!	1	A	Q		
0010	EOA	DC_2	"	2	B	R		
0011	EOM	DC_3	#	3	C	S		
0100	EOT	DC_4	$	4	D	T		
0101	WRU	ERR	%	5	E	U		
0110	RU	SYNC	&	6	F	V		
0111	BELL	LEM	.	7	G	W		
1000	FE_0	S_0	(	8	H	X		
1001	HT / SK	S_1	)	9	I	Y		
1010	LF	S_2	*	:	J	Z		
1011	V_{tab}	S_3	+	;	K	[		
1100	FF	S_4	,	<	L	\		ACK
1101	CR	S_5	–	=	M	]		②
1110	SO	S_6	★	>	N	↑		ESC
1111	SI	S_7	/	?	O	←		DEL

Definitions of control abbreviations:

ACK	Acknowledge	LEM	Logical end of media
BELL	Audible signal	LF	Line feed
CR	Carriage return	RU	"Are you. . .?"
DC_0 DC_4	Device control	SK	Skip
DEL	Delete idle	SI	Shift in
EOA	End of address	SO	Shift out
EOM	End of message	S_0 S_7	Separator (space)
EOT	End of transmission	SOM	Start of message
ERR	Error	V_{tab}	Vertical tabulation
ESC	Escape	WRU	"Who are you?"
FE	Format effector	②	Unassigned control
FF	Form feed	SYNC	Synchronous idle
HT	Horizontal tabulation		

Example of code format:

B_7 B_1

1000100 is the code for D

three-bit group   four-bit group

Figure 5-16. ASCII code

DIGITAL SYSTEM OPERATIONAL STATES

Digital systems require a precise definition of operational states or conditions in order to be useful. In actual circuit design applications, binary signals are considered to be far superior to those of the octal, decimal, or hexadecimal systems just discussed. In practice, binary signals can be processed very easily through electronic circuitry because they can be represented by two stable states of operation. These states can be easily defined as on or off, 1 or 0, up or down, voltage or no voltage, right or left, or any of the other two-condition designations. There must be no in-between step or condition. These states must be decidedly different and easily distinguished.

The symbols used to define the operational state of a binary system are very important. In positive binary logic, such things as voltage, on, true, or a letter designation such as A are used to denote the 1 operational state. No voltage, off, false, or the letter A are commonly used to denote the alternate, or 0, condition. An operating system can be set to either state, where it will remain until something causes it to change conditions.

Any device that can be set in one of two operational states or conditions by an outside signal is said to be *bistable*. Switches, relays, transistors, diodes, and ICs are commonly used examples. In a strict sense, a bistable device has the capability of storing one binary digit or bit of information. By employing a number of these devices, it is possible to build an electronic circuit that will make decisions based on the applied input signals. The output of such a circuit is, therefore, a decision based on the operational conditions of the input. Since this application of a bistable device makes logical decisions, it is commonly called a binary logic circuit, or simply a logic circuit.

There are two basic types of logic circuits in a digital system. One type of logic circuit is designed to make decisions. It has data applied to its input and produces an output that coincides with a prescribed combination of rules. Electronic decisions are made with logic gates. Memory is the other type of logic circuit. Memory circuits store binary data. These data can be stored and retrieved from memory when the need arises. Special ICs are used to achieve the memory function of a digital system. Memory is a primary function of a digital system. Performance is largely dependent on the capacity of a system's memory.

SUMMARY

A digital system has a source of operating energy, a path, control, a load device, and an indicator. The digital system also has an internal signal that is generated and processed by the circuitry. The path of electrical energy is usually a printed circuit board. Control is achieved by logic gates made of bipolar transistors, MOSFETs, and diodes. The load of the dc energy source is all the components that receive operational energy in the system. The indicator of the system can be a digital display or a cathode-ray tube.

A decimal numbering system has 10 individual values or symbols in its makeup. Any number from 0 to 9 can be used in a place location. The next location to the left of the units place is the 10s position. Each succeeding place to the left has a value that is 10 times that of the preceding place. Electronically the decimal system is rather difficult to use. Each number would require a specific value to distinguish it from the others. Number detection would also require some unique method of distinguishing each value from the others.

Nearly all digital systems use binary numbers in their operation. A binary number has 2 as its base. The largest number that can be expressed at a place location is 1. Only is and Os are used in a binary system. These are called binary digits, or bits. The operational basis of a binary system is very similar to that of the decimal system. The first place to the left of the binary point is the is place. Place values are expressed as powers of 2.

Numbers are frequently converted from one system to another. Conversion of a decimal number to a binary number is achieved by repetitive steps of division by the number 2. When there is no remainder, a 0 is recorded. A 1 is recorded when there is a remainder. The division process continues until a quotient of 0 is obtained. The binary equivalent is the accumulation of the recorded remainders. Binary-to-decimal conversion is achieved by starting at the binary point to find the decimal value for each binary place. Then the place-value assignments are added to find the decimal number equivalent.

A binary-coded decimal number is used to indicate large binary numbers. This numbering system is a weighted code. Four binary digits are used to represent each decimal digit. The largest digit to be displayed by any binary group of the BCD number is 9. The four-binary-group number is separated from other groups by a dash when it is displayed.

Octal, or base 8, numbering systems are used to process large numbers through a digital system. The largest digit used by the system is the number 7. The place values of digits are powers of 8. The process of converting an octal number to an equivalent binary number is similar to BCD conversion. The octal number is first divided into discrete digits according to place value. Each octal digit is then converted into an equivalent binary number using only three digits. Converting a decimal number to an octal number is a process of repetitive division by the number 8. The quotient is transferred for the next division step and the remainder is recorded. The accumulated remainders represent the octal equivalent of the decimal number. Converting a binary number to an octal number is achieved by dividing the binary number into groups of three digits starting at the left of the octal point. Each binary number group is then converted into an equivalent octal number. These numbers are then combined to form the octal number.

Hexadecimal numbers are used in digital systems to process large numbers. The base of this system is 16. The largest value displayed at a specific place is 15, which is displayed by digits 0 through 9 and the letters A, B, C, D, E, and F. The place values from the left of the hexadecimal point are powers of 16. Conversion of a hexadecimal number to a decimal number is similar to the other conversion processes. The hexadecimal number is recorded and the place values or powers of the base are then positioned under each digit. The values are then multiplied to indicate discrete place values. Conversion of decimal to hexadecimal is achieved by repetitive division. The quotient is transferred for the next division step and the remainder is recorded. The combined remainders form the hexadecimal number. Conversion of a binary number to hexadecimal is achieved by dividing the number into groups of four digits starting at the hexadecimal point. Each binary group is then changed into an equivalent hexadecimal value.

Binary codes have been developed to interface a digital system between binary data and alphanumeric data. The BCD code was developed to display decimal numbers in groups of four binary bits of data that could be easily defined. The parity code is an error-checking code. It has a binary bit that is used to detect errors. The odd parity code generates a 0 parity bit when the number of 1s is odd. A 1 is generated when the number of 1s is even. The even parity code is just the opposite of the odd code. The Gray code is a method of displaying binary data with only one bit of data between each number. It is used to encode data such as shaft position of a motor or physical location of moving parts. Alphanumeric codes are used to represent numbers, letters, and punctuation marks as binary data. The ASCII code is widely used for this operation in a digital system. The ASCII code is a 7-bit code. It has a three-bit group for the column-select and a four-bit group for the row-select function. The combined 7-bit number represents a specific letter, number, or punctuation mar.

Chapter 6

Binary Addition and Subtraction

INTRODUCTION

Digital systems are called on to perform a number of arithmetic functions in their operations. These functions deal with numbers in a variety of different forms. As a rule, the circuitry of a digital system must deal with binary numbers. The internal circuitry must manipulate the data very quickly and accurately.

The subject of digital arithmetic can be very complex when all of the methods of computation and the theory behind it are understood. Fortunately, this level of understanding is not required by technicians until they become experienced system programmers. In this unit, we concentrate on those basic principles needed to understand how digital systems perform basic arithmetic operations. We first discuss arithmetic operations on binary numbers using pencil and paper. We then examine logic circuits that perform these operations in a digital system. As a rule, most arithmetic operations can be achieved with integrated circuits. Special arithmetic ICs are readily available as off-the-shelf items. A person working with digital systems must be familiar with the operation of these devices in order to see how a system functions electronically.

ADDITION

Addition is a mathematical operation in which two or more numbers are combined together to obtain a simple equivalent quantity. Addition of integer A to integer B, or $A + B$, is the process of advancing integer A, B times. Addition is a process of memorizing a list of number combinations. In practice, it is extremely difficult to make a list of all the integer combinations needed to accomplish addition. A list of the base number combinations is however, very helpful in seeing how addition is accomplished. Addition is basically a counting operation. The counting method of addition is rather slow and somewhat awkward to handle when large numbers are involved. It is generally accomplished by memorizing an addition table that serves as a list of representative small-number combinations. This table also shows some of the rules that are used to perform the addition operation. After repeated use of the addition operation, the table is not needed and we can add numbers together through memorization of the number combinations and rules.

To add in other number systems, we simply need to memorize a table of integer combinations of the numbering system base. The basic rules of addition are similar for all positive bases. Figure 6-1 shows a simplified addition table for decimal numbers up to 9. The number in each position in the first row represents a simple integer value. This number can be called the *addend* of the addition operation. The numbers in the leftmost column represent a second set of integer values, also called the *addends.* Addition of the values in the top row to those in the left column results in the sums of these values. To use the table, simply find an integer in the top row. In this case let us use the number 3. An integer from

Operation   Addend

+	0	1	2	3	4	5	6	7	8	9
0	0	1	2	3	4	5	6	7	8	9
1	1	2	3	4	5	6	7	8	9	10
2	2	3	4	5	6	7	8	9	10	11
3	3	4	5	6	7	8	9	10	11	12
4	4	5	6	7	8	9	10	11	12	13
5	5	6	7	8	9	10	11	12	13	14
6	6	7	8	9	10	11	12	13	14	15
7	7	8	9	10	11	12	13	14	15	16
8	8	9	10	11	12	13	14	15	16	17
9	9	10	11	12	13	14	15	16	17	18

Addend (left of table)   Sum (right of table)

Sum

Figure 6-1. Decimal addition table

the left column to be added to 3 is then found. Assume that the addend is 5. Locate 5 in the left column. Trace across the numbers of the 5 horizontal row until it intersects with the 3 column of the addend. The intersection of 5 horizontally with 3 in the vertical column shows the sum to be 8. This indicates that $5 + 3 = 8$. The table shows a variety of different integer combinations for numbers up to 9. One important rule of addition is shown by the table when the sum of an integer combination is larger than 9. When this occurs a carry is generated. This keeps the place value of the system in proper prospective. Note that $5 + 5 = 10$. In this example, the sum has two places. The right side of the two placed number is the units value and the second number represents the 10s place. In a decimal number system, the integer value at any place cannot exceed a value of 9. When it does, an integer in the 10s place must be used to represent the number. Larger value decimal numbers are a combination of multiple place values in composite form.

Binary Addition

The addition of binary numbers is similar in nearly all respects to the addition of decimal numbers. A binary addition table is shown in Figure 6-2. This table is a simplification of the decimal table. It only has four possible alternatives. The procedure for using the table is similar to that of the decimal table. A binary addition table only has two integers. Only one binary digit, or bit, is needed to indicate the sum of three of these binary values. The sum of $1 + 1$ necessitates more than one place to be displayed. In this case, $1 + 1 = 10$. A carry is generated and transferred to the second, or 2s, place. Look at the binary table and try adding the designated number combinations. These are $0 + 0 = 0$, $0 + 1 = 1$, $1 + 0 = 1$, and $1 + 1 = 10$. The binary table is easily memorized because of its simplicity and brevity. It is very important in electronic arithmetic because it has only two digits.

The binary numbering system was discussed in the first unit of this course. Binary numbers are used to represent different logic functions and to define counting operations. In this unit we are concerned with arithmetic operations that use binary numbers. We must be able to perform binary addition in order to see how a digital system accomplishes this operation electronically. Some sample problems are shown at this time to assist you in using the rules of binary addition. The problems are presented in a standard addition format. This is similar to the addition of decimal values.

There are four basic rules to follow in binary addition. These were expressed in the addition table of Figure 6-2. The standard addition for-

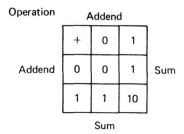

Figure 6-2. Binary addition table

	Rule 1	Rule 2	Rule 3	Rule 4
Addend	0	0	1	1
Addend	+ 0	+ 1	+ 0	+ 1
Sum	0	1	1	1 0

Carry (↑ under Rule 4)

Figure 6-3. Binary addition rules

mat of these rules is shown in Figure 6-3. Note the location of the addends, and sum. The first three rules are the same as decimal addition. Rule 4 is somewhat unique. It shows that 1 plus 1 equals 0 with a carry of 1. Remember that $2_{10} = 10_2$. Therefore, $1 + 1 = 10$ in binary numbers.

When large binary numbers are to be added, the same basic rules apply. The carry operation generally causes some concern. The carry must be added to the next order of numbers. In decimal addition the carry is generally added to the next column by a notation. Some people simply remember the carry value or place it into their own memories. Others may add the carry to the next column of numbers by penciling in the value. Regardless of the procedure, the carry function must be taken into account. Figure 6-4 shows some problems that use the basic rules of binary addition. The decimal equivalent of each number is also shown for comparison of the addition process. Work through the problems following the rules of binary and decimal addition. Problems 1 and 2 employ the first three rules of binary addition. Problems 3 through 5 utilize all four binary addition rules. Note the manipulation of the carry. In problem 6, the carry function is somewhat more complex. Three 1s must be added in one position. The additional 1 comes from the carry of the preceding column. Theoretically, this operation necessitates a subtotal listing. It can be done as listed or can

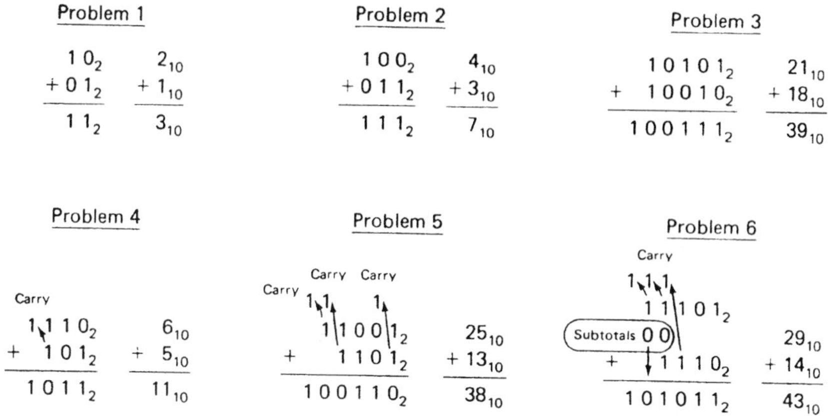

Problem 1		Problem 2		Problem 3	
$1\ 0_2$	2_{10}	$1\ 0\ 0_2$	4_{10}	$1\ 0\ 1\ 0\ 1_2$	21_{10}
$+\ 0\ 1_2$	$+\ 1_{10}$	$+\ 0\ 1\ 1_2$	$+\ 3_{10}$	$+\ \ \ 1\ 0\ 0\ 1\ 0_2$	$+\ 18_{10}$
$1\ 1_2$	3_{10}	$1\ 1\ 1_2$	7_{10}	$1\ 0\ 0\ 1\ 1\ 1_2$	39_{10}

Problem 4		Problem 5		Problem 6	
				Carry	
		Carry Carry		$1\ 1\ 1$	
Carry		Carry $1\ 1$ 1		$1\ 1\ 1\ 0\ 1_2$	
$1\ 1\ 1\ 0_2$	6_{10}	$1\ 1\ 0\ 0_2$	25_{10}	Subtotals $0\ 0$	29_{10}
$+\ \ \ 1\ 0\ 1_2$	$+\ \ 5_{10}$	$+\ \ \ 1\ 1\ 0\ 1_2$	$+\ 13_{10}$	$+\ \ \ 1\ 1\ 1\ 0_2$	$+\ 14_{10}$
$1\ 0\ 1\ 1_2$	11_{10}	$1\ 0\ 0\ 1\ 1\ 0_2$	38_{10}	$1\ 0\ 1\ 0\ 1\ 1_2$	43_{10}

Figure 6-4. Binary addition problems

be accomplished by memory. The sum of this column is 1 because $1 + 1 = 0 + 1 = 1$ and a carry is generated to the next column. Note this particular operation.

Binary Addition Circuits

Logic gates are used in digital systems to accomplish the arithmetic operation of binary addition. Addition of two binary numbers necessitates a two-input logic gate. The logic gate needed for this operation is somewhat unique. It, for example, must accept two inputs and develop an output that is high (1) only when the inputs are different. This logic function is not included in the basic logic gate functions of AND, OR, NOT, NAND, and NOR. A special logic gate is needed for this operation. The exclusive-OR gate is used to achieve binary addition. This gate is slightly different from the basic OR gate or the inclusive-OR function. An inclusive-OR gate adds integers together and is described as OR addition. This operation is defined by the expression $A + B = C$. ORing two integers together produces a 1 output when either or both inputs are 1. An OR gate, however, does not achieve binary addition. You may recall that the rules of binary addition are $0 + 0 = 0$, $0 + 1 = 1$, $1 + 0 = 1$, and $1 + 1 = 10$. A high (1) appears in the output only when the two inputs are different. Truth tables showing the function of an inclusive-OR gate and the exclusive-OR gate are shown in Figure 6-5. Note the logic symbols and the Boolean expressions for each gate.

The Boolean expression of a truth table can be written by a technique learned earlier. The output conditions of the truth table containing a 1 are

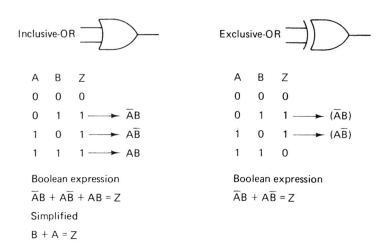

Figure 6-5. Inclusivie-OR and exclusive-OR gate comparisons

translated into input states. The inclusive-OR gate has three outputs containing a 1. These outputs are expressed as the product of the inputs. The complete expression is the sum of products. For the inclusive-OR gate this is $\bar{A}B + A\bar{B} + AB = Z$. This can be simplified to $A + B = Z$. For the exclusive-OR gate, the Boolean expression for the truth table is $\bar{A}B + A\bar{B} = Z$. Comparison of the two expressions shows a distinct difference between the inclusive- and exclusive-OR gates. The inclusive-OR gate achieves OR addition (+) of integers and the exclusive-OR gate achieves binary addition (Z).

An exclusive-OR gate can be implemented by several different logic gate combinations. The process of translating a Boolean expression into a logic gate circuit was discussed in Unit 5. Refer to the Boolean expression of the exclusive-OR gate. Design a logic gate circuit that will accomplish this operation. Evaluate the operation of the gate circuit by substituting the integer values of A and *B* for each step of the truth table. Does your gate circuit perform the binary addition operation?

Figure 6-6 shows several ways of implementing the exclusive-OR function with logic gates. See if your implementation circuit is included in this listing. As a rule, the IC family used in the construction of a logic gate usually dictates the gate circuitry employed by a chip. TTL uses the NAND gate as its fundamental building block. CMOS uses the NOT gate as its fundamental building block. It is common practice to build several exclusive-OR gates on a single IC chip. The 74LS86 has four X-OR gates

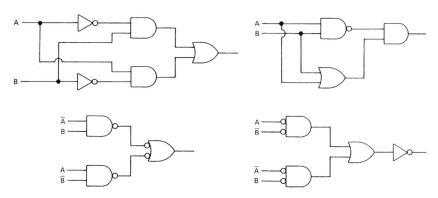

Figure 6-6. Exclusivie-OR gate implementation circuits

built on a 14-pin dual in-line package. The standard logic symbol of an exclusive-OR gate is used to identify the discrete gates of this package. The 74LS86 is a member of the TTL logic family.

Half-Adders

The exclusive-OR (X-OR) gate is commonly used in binary addition circuits. However, it can be used only to accomplish the first three rules of binary addition. An X-OR gate does not develop a carry when two 1s are added. Because of this limitation the exclusive-OR gate needs some additional circuitry to accomplish binary addition. The carry function, for example, occurs only when two 1s are applied to the input. An AND gate can be used to denote when this condition exists. By connecting an AND gate to the input of an exclusive-OR gate, a half-adder is formed. Figure 6-7 shows the gate circuitry of a half-adder. A half-adder has two inputs, a sum output and a carry output. It is used to add first order binary numbers. The truth table shows the response of a half-adder. Since two outputs are produced by a half-adder, two Boolean expressions are included in its function. Note the Boolean expressions and truth table of the half-adder.

A half-adder is generally represented as a functional block in logic diagrams. The logic symbol of a half-adder is a square surrounding the letters H.A. The inputs are usually labeled A and B. The sum output is identified by the Greek letter sigma Σ. The carry output is identified by the letters C_o. Half-adders are used primarily to add the augend and addend of a first-order binary number. The addends can be reversed without affecting the operation. A half-adder has a limited number of applications in binary addition when used as an independent logic block. This limitation

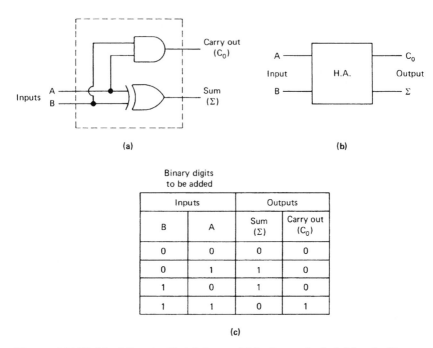

(a)   (b)

Binary digits
to be added

Inputs		Outputs	
B	A	Sum (Σ)	Carry out (C_0)
0	0	0	0
0	1	1	0
1	0	1	0
1	1	0	1

(c)

Figure 6-7. Half-adder, truth table, and block symbol: (a) logic diagram of a half-adder; (b) half-adder block symbol, (c) truth table.

is based on its ability to add only a single column of two numbers in the first order of a problem.

Full-Adders

Adding two first-order binary numbers electronically can be accomplished with a half-adder. The first column, or order, of the number is generally made up of only two digits. Second-order numbers, or the 2s column, may involve the addition of three digits.

A carry generated by the first-order numbers may appear as an input for the second-order number. A half-adder can not respond to three inputs. To add three digits at a time, a full-adder must be used.

A full-adder accepts three inputs and generates a sum and carry output. The full-adder accommodates two input bits and a carry input. In order to accomplish this operation, a full-adder may be formed by two half-adders and an OR gate. The composition of a full-adder is shown in Figure 6-8. Each half-adder of the circuit is surrounded by broken lines. Note the operation achieved by different parts of the full-adder. Each half-

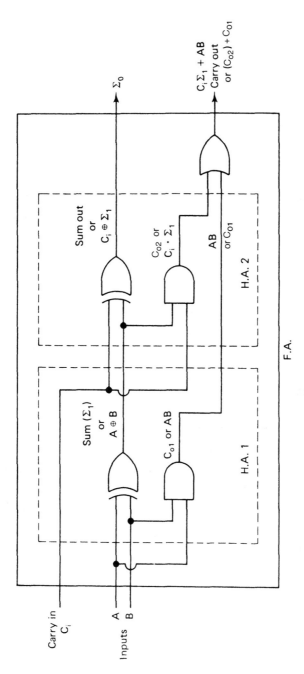

Figure 6-8. Full-adder composition

adder responds in a conventional manner. The sum output (Σ_1) of the first half-adder is $A \oplus B$. The carry out (C_{o1}) is AB. The sum output of the second half-adder (the composite output of the full-adder) is labeled out (Σ_o). The Σ_o of the full-adder equals the carry in (C_1) $\oplus$ Σ_1. The carry out (C_{O2}) of the second half-adder is $C_1 \Sigma_1$. The carry out (C.) of the composite full-adder is $C_{ol} + C_{o2}$. The solid line surrounding the composite adder denotes the internal structure of a full-adder.

A full adder is represented by a block surrounding the letters F.A. This symbol representation of the full-adder is shown in Figure 6-9. Note that the symbol has its inputs labeled A, B, and C_{in}. The outputs are labeled Σ_o and C_o.

The logic function of a full-adder is shown by the truth table of Figure 6-9. The inputs are indicated as A, B, and C_{in}. The outputs are labeled as the sum (Σ_o) and carry out (C_o). Try the addition sequence of the binary numbers listed in the truth table. The sum and carry out of the truth table should show the response of the adder. This accomplishes the fundamental rules of binary addition.

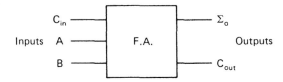

	Inputs		Outputs	
C_{in}	B	A	Σ_o	C_{out}
0	0	0	0	0
0	0	1	1	0
0	1	0	1	0
0	1	1	0	1
1	0	0	1	0
1	0	1	0	1
1	1	0	0	1
1	1	1	1	1

Figure 6-9. Full-adder symbol and truth table

In practice, four full-adders are housed in a common dual in-line package IC. A representative 4-bit full-adder is the 74LS283. This IC is identified as 4-bit fast carry full-adder. Each adder is identified by a numbering sequence such as A_1, B_1, Σ_1, and C_{in}. The carry output of full-adder number 1 is automatically transferred to the input of full-adder number 2. The same holds true for the adders 3 and 4. The final output of adder number 4 appears as Σ_4 and C_{out}. The number of bits handled by this type of adder can be increased in multiples of four by adding additional ICs. The C_0 of the first IC serves as the C_{in} of the next succeeding adder configuration.

SUBTRACTION

Subtraction is a mathematical operation in which one integer or number is deducted from another to obtain an equivalent quantity. Subtraction reduces integer A, B times.

When subtraction is performed, the number from which another number is to be subtracted is called the minuend. The number subtracted from the minuend is the subtrahend. The remainder, or difference, is the resultant of the subtrahend being subtracted from the minuend. Some sample subtraction problems involving decimal numbers are shown in Figure 6-10. Problem 1 shows the subtraction of single-place decimal numbers. Problem 2 shows the subtraction of two-place numbers. Remember that two-place numbers represent the number of 10s and is in a decimal system. Problem 3 has three-place numbers. These represent 1s, 10s, and 100s. Problem 4 has four-place values. These represent ls, 10s, 100s, and 1000s.

Figure 6-11 shows a simplified subtraction table for decimal numbers up to 10. Each number in the top row represents an integer, the minuend. These numbers start with the largest value (10) and decrease in value to 0. The numbers in the leftmost column represent a second set of integer

Problem 1		Problem 2	Problem 3	Problem 4
5	Minuend	68	294	3167
− 3	Subtrahend	− 21	− 183	− 2154
2	Difference	47	111	1013

Figure 6-10. Decimal subtraction

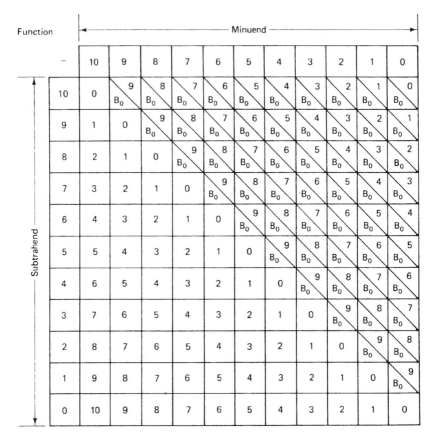

Figure 6-11. Subtraction table

values, the subtrahends. Subtraction of the numbers in the left column (subtrahend) from those in the top row (minuend) produces the differences. To use the table, simply find an integer in the top row. In this case, let us use the number 5. Then find a subtrahend in the left column. Assume that the subtrahend is 3. Locate 3 in the left column. Then trace across the numbers in the 3 horizontal row until it intersects with the integer in 5 column. This shows that the difference between 5 and 3 is 2 (5 – 3 = 2).

The subtraction table of Figure 6-11 shows a variety of different integer combinations for numbers up to 10. An important rule of subtraction is also shown by the table when the minuend is smaller than the subtrahend. In order to subtract these integers, a borrow must be initiated. This assumes that the minuend has a place value greater than one integer. To show this application, let us subtract the single place-value number 9 from the two-place number 10. Locate the 10 column in the top horizontal row. Then locate a subtrahend of 9. Tracing across the 9 row to the 10 column shows that 10 − 9 = 1. In this case, 9 is greater than 0. A borrow from the second place value is needed to increase the value of 0. Therefore, 0 becomes 10. 9 subtracted from 10 is 1, or 10 − 9 = 1. A borrow is not indicated in the difference box of the table for this problem because the number 10 has two places. For single integer values, a diagonal line divides the difference box. The notation B. indicates that a borrow is needed to convert the integer to a two-place number. The difference is shown above the diagonal line. Half of the table shows the difference being accomplished by the borrow function. Work through the table to see how these number combinations are achieved.

In practice, a person does not perform the subtraction operation by using a table like the one shown in Figure 6-11. The different number combinations of the table are memorized and are easily recognized after repeated use. The basic operational rules of subtraction are, however, very important. All positive-base numbering systems use the same basic rules for subtraction. We simply need to memorize a table of integer combinations of the numbering system to use in subtracting.

Binary Subtraction

The subtraction of binary numbers is similar in nearly all respects to the subtraction of decimal numbers. A binary subtraction table is shown in Figure 6-12. This table is a simplification of the decimal table. It only has four possible alternatives. The procedure for using the table is similar to that of the decimal table. A binary subtraction table only has two integers, 1 and 0. Only one bit is needed to indicate the difference

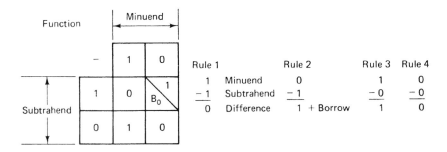

Figure 6-12. Binary subtraction rules and table

in three of these binary values. Rule 2 shows us that $0 - 1 = 1$, and a borrow must be taken from the second, or 2s, place. This is similar to the subtraction of decimal numbers when the place value of the minuend is smaller than the subtrahend. Look at the table and the subtraction rules of the other binary numbers. These are $1 - 1 = 0$, $0 - 1 = 1$ and a borrow, $1 - 0 = 1$, and $0 - 0 = 0$. These rules are easily memorized because of their simplicity. Binary subtraction is a very important electronic arithmetic process.

When binary numbers with more than two place values are to be subtracted, the same basic rules apply. The borrow operation generally causes some confusion. The borrow must be deducted from the next highest place. In decimal subtraction the borrow is simply deducted from the next column by notation. Some people simply remember the borrow has taken place and others pencil in the borrow. Regardless of the procedure, the borrow function must be taken into account. Figure 6-13 shows some problems that use the basic rules of binary subtraction. A decimal equivalent of each number is also shown for comparison. Work through the problems following the rules of binary and decimal subtraction. The first three problems do not require borrowing. Problems 4 through 6 utilize all four binary subtraction rules. Note that the borrow operation is penciled in to show when it is used. Problem 4 does not require a borrow until the numbers in the third place are subtracted: 0 - 1 = 1 and a borrow from the fourth-place number. Borrowing a 1 from this place changes its value to 0. As a result, the process ends at this position. Problem 5 uses the borrow operation in the first order, or 1 s position: 0

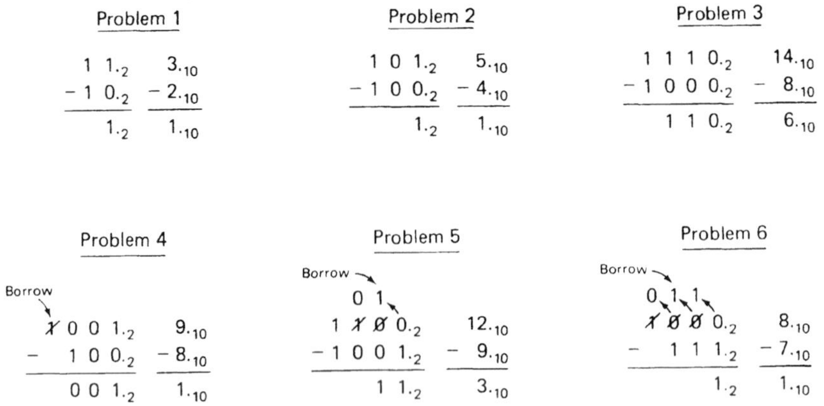

Problem 1		Problem 2		Problem 3	
$1\ 1._2$	$3._{10}$	$1\ 0\ 1._2$	$5._{10}$	$1\ 1\ 1\ 0._2$	$14._{10}$
$-\ 1\ 0._2$	$-\ 2._{10}$	$-\ 1\ 0\ 0._2$	$-\ 4._{10}$	$-\ 1\ 0\ 0\ 0._2$	$-\ 8._{10}$
$1._2$	$1._{10}$	$1._2$	$1._{10}$	$1\ 1\ 0._2$	$6._{10}$

Problem 4		Problem 5		Problem 6	
Borrow		Borrow $\searrow$		Borrow $\searrow$	
		$0\ 1\searrow$		$0\searrow 1\searrow 1\searrow$	
$\cancel{1}\ 0\ 0\ 1._2$	$9._{10}$	$1\ \cancel{1}\ \cancel{0}\ 0._2$	$12._{10}$	$\cancel{1}\ \cancel{0}\ \cancel{0}\ 0._2$	$8._{10}$
$-\quad 1\ 0\ 0._2$	$-\ 8._{10}$	$-\ 1\ 0\ 0\ 1._2$	$-\ 9._{10}$	$-\quad 1\ 1\ 1._2$	$-\ 7._{10}$
$0\ 0\ 1._2$	$1._{10}$	$1\ 1._2$	$3._{10}$	$1._2$	$1._{10}$

Figure 6-13. Binary subtraction problems

- 1 = 1 and a borrow. The borrow must come from the second place, or 2s position. The 0 at this position is changed into a 1. Note the integer value change. Since the second place value was a 0, it needs a borrow from the third place. The 1 of the third place is changed to a 0 after the borrow. The subtraction process continues with 1 - 0 = 1 in the 2s place and 1 - 0 = 1 in the 4s place. In the 8s place, 1 - 1 = 0 completes the operation. Problem 6 is similar, with a borrow being generated in the 1s place. This borrow changes the succeeding minuend values from 0s to 1s. The subtraction process now involves the new minuend values. 1 - 1 = 0, 1 - 1 = 0, and 0 - 0 = 0 are all understood to be 0 and are not noted in the remainder, or difference. Practice the operational steps of this problem to see how the result is achieved. You may want to form some of your own problems to practice the subtraction process. Practice is generally needed to become proficient in the subtraction of binary numbers.

Binary Subtraction Circuits

Subtraction can be accomplished with a two-input logic gate. The rules of binary subtraction are shown in the table of Figure 6-12. Note that these rules show a difference of 1 only when the minuend and subtrahend are different. When they are the same, the difference is low, or 0. This operation can be accomplished with an exclusive-OR gate. The Boolean expression of an exclusive-OR gate is $\bar{A}B + A\bar{B} = C$. An exclusive-OR gate can be implemented several different ways with logic gates. The X-OR gate can be used to determine the difference only in a column of first-order numbers. In order to accomplish all the rules of binary subtraction, a borrow must be implemented for one rule. This necessitates some additional gates connected with the exclusive-OR gate. This combination is called a half-subtractor. The input of a half-subtractor is the minuend and subtrahend. The output is the difference and borrow.

Figure 6-14 shows the logic circuitry and truth table of a half-subtractor. This circuit deducts the subtrahend from the minuend with an exclusive-OR gate. The borrow function is accomplished with an AND gate. The minuend (input *A)* of the borrow circuit is first applied to a NOT gate. The output of the AND gate is $\bar{A}B$. This represents the 0 – 1 = 1 condition in which a borrow is involved. A half-subtractor can be used to accomplish subtraction only of first-order, or least-significant number, combinations. To accomplish higher-order number combinations a full-subtractor is needed.

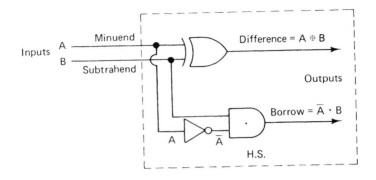

Truth table

Minuend A	Subtrahend B	Difference	Borrow
0	0	0	0
0	1	1	1
1	0	1	0
1	1	0	0

Figure 6-14. Half-subtractor and truth table

Full-Subtractors

A full-subtractor accepts three inputs and generates difference and borrow outputs. The three inputs are the minuend, subtrahend, and a borrow-input. In order to accomplish this arithmetic operation, a full-subtractor is formed with two half-subtractors and an OR gate. The composition of a full-subtractor is shown by the simplified block diagram of Figure 6-15.

The logic gate structure of the two half-subtractors of a full-subtractor is similar to that of Figure 6-14. Two half-subtractors are interconnected to produce a borrow output (B_o) and a difference output (D_o). The half-subtractors respond in a conventional manner. The difference output of the first half-subtractor is $A \oplus B$. This produces difference output 1, or D_1. The borrow developed by half-subtractor 1 is $\bar{A}B$. This is called B_1. B_1 is then applied to an OR gate and D_1 is applied to the input of half-subtractor 2. Half-subtractor 2 has D_1 as one of its inputs and the borrow in (C_{in}) as the other input. The difference output (D_o) is $A \oplus B \oplus C_{in}$. The borrow output (B_o) has B_1 and B_2 applied to its inputs. B_o is equal to $\bar{A}B + (\overline{A \oplus B})C_{in}$. The compos-

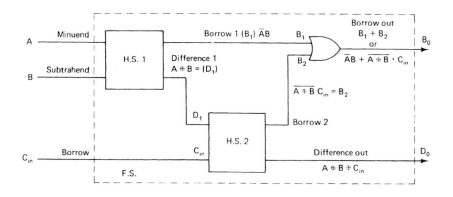

Truth table

Borrow in C	Subtrahend B	Minuend A	Difference out D_0	Borrow out B_0
0	0	0	0	0
0	0	1	1	0
0	1	0	1	1
0	1	1	0	0
1	0	0	1	1
1	0	1	0	0
1	1	0	0	1
1	1	1	1	1

Figure 6-15. Full-subtractor and truth table

ite full-subtractor, therefore, has inputs of A (minuend), B (subtrahend), and C_{in} (borrow in). The outputs are B_0 (Borrow out) and D_0 (difference out).

A binary subtractor is commonly treated as a functional block just like an AND, OR, or NOT gate. In fact, subtraction and addition can be accomplished with the same functional block with some modification. The exclusive-OR function, for example, is the same for both subtraction and addition. There is, however, a difference between the borrow and carry functions. The common usage of microprocessors in digital systems has diminished the importance of functional block subtractors and adders. Adders are available, but subtractors are not. Subtraction can be achieved with adders by using another function called complementation. In fact, subtraction, multiplication, and division can all be accomplished by the addition process. This operation is demonstrated by the use of a functional block called the arithmetic logic unit, or ALU.

Parallel Subtractors

Half- and full-subtractors are generally wired together to perform as a parallel subtractor. The half-subtractor responds to first-order numbers of the problem and succeeding full-subtractors respond to higher-order numbers. A parallel binary subtractor is shown in Figure 6-16. Note that this subtractor will accommodate a 4-bit binary number. The first bit is applied to a half-subtractor. This part of the subtractor accommodates only the minuend and the subtrahend as its input. Its output is the difference and the borrow. The borrow is automatically transferred to the input of the next most significant place. This part of the circuit necessitates the use of a full-subtractor. Twos place data are applied to input A_1, and B_1 externally. A_1 is the minuend and B_1 is the subtrahend. The output is the 2s place difference and a borrow. The borrow is automatically coupled to the 4s place. This requires a full-subtractor to accommodate the three inputs. The circuitry is duplicated for the 8s place data. A subtractor of this type can accommodate only a 4-bit binary number. The number designations for this subtractor are identified as $A_0 - B_0$, $A_1 - B_1$, $A_2 - B_2$, and $A_3 - B_3$. Subtractors are generally not available on ICs today. A subtractor can be built with discrete gates that will accomplish this function.

Subtracting with Adders

Binary subtraction is generally not performed in the manner just described. Subtraction by this method would necessitate some independent logic circuits that could be connected to form half-subtractors and full-subtractors. As a rule, this complicates the circuitry of a calculator. To simplify the circuitry, a more universal arithmetic device is used. Subtraction, for example, can be achieved in this device by addition. An adder can, therefore, be used to accomplish both arithmetic operations with the same circuitry. This is achieved by playing a few tricks on the adder. The subtrahend, for example, must be complemented before it is added. To complement a number we simply invert each bit of data. Each 1 is changed to 0 and each 0 is changed to 1. The minuend and complemented subtrahend are then added together to produce a difference. The most-significant bit of data must be removed from the difference and added to the remainder. This is generally called an end-around carry. This subtraction operation is called the 1s complement method.

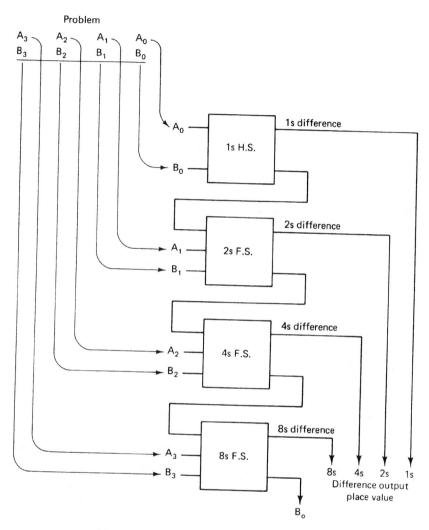

Figure 6-16. Four-bit parallel subtractor

1s Complement Subtraction

The 1s complement method of subtraction is shown in Figure 6-17. This method of subtraction is first shown in decimal form. Note that the minuend is the decimal 10 and the subtrahend is the decimal 4. The difference is a decimal value of 6. Subtraction of the equivalent binary numbers is also shown. The minuend is 1010_2 and the subtrahend is 0100_2. This produces a difference of 110_2. Subtraction by the 1s complement meth-

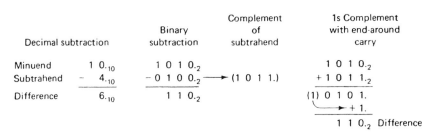

Figure 6-17. Subtraction by 1s complement with end-around carry

od is also shown in Figure 6-17. This method of subtraction necessitates complementing the subtrahend. Note that 0100_2 is changed to 1011_2. The complemented subtrahend is then added to the original minuend. This produces a sum of 10101_2. The most significant bit of this number is a 1. It is carried around to the least-significant column and added. This is the end-around carry operation. Note that the sum of $0101_2 + 1_2$ is 110_2. This is the binary equivalent of the decimal 6. Hence, $1010_2 + 1011_2 + 1_2 = 110_2$. This achieves the same results as $1010_2 - 0100_2 = 110_2$, or $1010_{10} - 4_{10} = 6_{10}$.

Figure 6-18 shows the circuitry of a 4-bit is complement subtractor. Note that the circuit is made up of four full-adders that are used to perform the subtraction operation. The inverters connected to inputs B_0, B_1, B_2, and B_3 complement the subtrahend. The complemented input is then added to the minuend, which is labeled A_0, A_1, A_2, and A_3. The end-around carry operation is achieved by connecting the carry output (C_o) of the 8s column back to the carry input (C_{in}) of the 1s adder. The sum output (Σ) of each adder represents the difference in binary numbers. This is in binary order and is expressed as 1s, 2s, 4s, and 8s. This subtractor can accommodate only 4-bit binary numbers. Larger numbers can be accommodated by connecting additional adders to the circuit. The carry out (C_o) of the 8s adder would be connected to the carry input (C_{in}) of the next series of adders. Extension of the circuit is usually accomplished 4 bits at a time. The carry out (C_o) of the last adder would be returned to the carry input (C_{in}) of the 1s adder. This type of circuit can be extended to accommodate any binary number.

2s Complement Subtraction

Computers generally use the 2s complement method of subtraction for binary numbers. It is easier to accomplish and uses less-complicated circuitry. The 2s complement method of subtraction is very similar to the

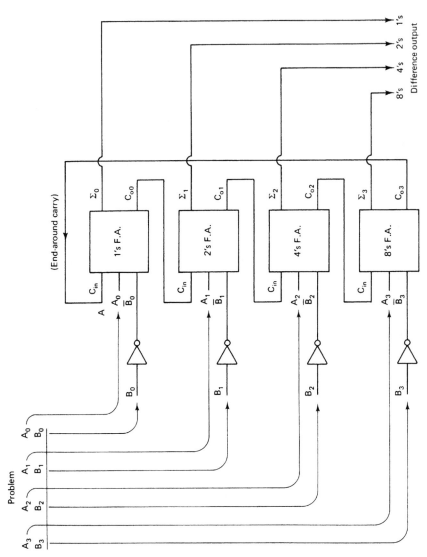

Figure 6-18. A four-bit subtractor using adders

1s complement. First, find the 1s complement of the subtrahend. Then add a 1 to it. This produces a 2s complement. The 2s complement subtrahend is then added to the minuend. The last, or most-significant carry, is dropped. The resulting sum is the difference.

Figure 6-19 shows more examples of subtraction by the 2s complement method. Each problem is shown in decimal form, conventional binary subtraction, and 2s complement form. Study the examples to become familiar with this operation. In problem 1 the decimal equivalent is $7 - 3 = 4$. The binary equivalent is $111 - 11 = 100$. The 2s complement method necessitates some changes. The subtrahend of the problem must first be complemented. This is the ls complement of 011. A 1 added to the 1s complement changes it to a 2s complement. The 2s complement is $100 + 1 = 101$. The minuend (111) is then added to the 2s complement subtrahend (101). The initial sum is 1100. Dropping the last carry, the final answer then becomes 100. This represents the difference of the problem. In effect, $111 + 101 = 1100$, for a corrected difference of 100.

The second problem in Figure 6-19 involves larger numbers. In this case $12_{10} - 5_{10} = 7_{10}$. The binary subtraction equivalent is $1100 - 101 = 111$. In the 2s complement version of the problem, the subtrahend first is converted to a 1s complement. A 1 added to the 1s complement changes it to

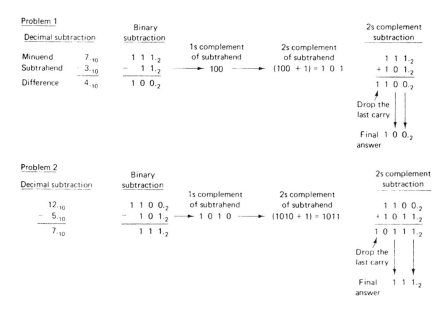

Figure 6-19. Twos complement subtraction

a 2s complement. The 2s complement is 1010 + 1, or 1011. The minuend (1100) is then added to the 2s complement subtrahend (1100). The initial sum is 10111. Dropping the last carry the final answer becomes 111. This represents the difference of the problem. Thus, 1100 + 1011 = 10111, for a corrected difference of 111_2.

A circuit that utilizes 2s complement numbers to achieve subtraction is shown in Figure 6-20. This circuit subtracts 4-bits of data. Four full-

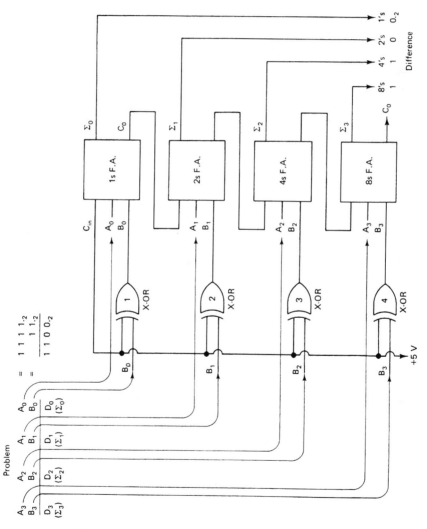

Figure 6-20. Subtraction using 2's complement

adders are needed to perform the operation. The minuend numbers A_3, A_2, A_1, and A_0 are applied directly to the A input of each full-adder. The subtrahend numbers are identified as B_3, B_2, B_1, and B_0. Each subtrahend number value is first applied to one input of an exclusive-OR gate. The alternate input of each X-OR gate is made high by connecting it to the + 5 V source. This causes each exclusive-OR gate to invert the data applied to its input. This data is then applied to the B input of each full-adder. The C_{in} of the ls full-adder is also held at a high level by attaching it to the + 5 V. The response of each exclusive-OR gate and the C_{in} of the 1s full-adder develops the 2s complement of the subtrahend. Essentially, this inverts the subtrahend input and adds a 1 to the ls complemented value. This changes the subtrahend (0011) to 1100 + 1 or 1101, which is the 2s complement of the original binary subtrahend. The A input of each full-adder then accepts the minuend (1111). The addition operation is then performed. The sum output is 1100_2. The last carry, C_o, is not recognized and is deleted from the final answer. The resulting sum (1100) is actually the difference in the minuend and the subtrahend. This shows that $1111_2 - 11_2 = 1100_2$.

ARITHMETIC LOGIC UNITS

A digital system such as a computer, calculator, robot, or programmable controller is called on to perform a variety of arithmetic operations. These range from basic arithmetic functions to mathematical operations that combine several functions in a complex expression. The logic circuitry for each arithmetic operation is quite unique. If each function required a separate logic circuit to perform its operation, the system would be rather massive. Today, digital systems employ ICs that are capable of doing several arithmetic operations in a single unit. The arithmetic logic unit, or ALU, was developed to perform this type of operation. The circuitry of the ALU dictates the arithmetic operations that it can accomplish. Some ALUs are capable of a few arithmetic operations such as addition, subtraction, and multiplication. Remember that addition can be used to subtract, multiply and divide digital numbers. Other ALUs respond to the basic arithmetic operations plus compare magnitudes, perform a variety of two-variable logic operations, and develop look-ahead carry signals for high-speed processing. These ALUs generally have their own internal registers for memory. Several ALUs of this type are available as off-the-shelf items. These chips generally manipulate data very quickly. An ALU may also be included as a functional part of a computer or a microprocessor chip. This type of ALU generally requires some support from other chips. The ALU must be interfaced with the system and may have some additional registers. This application generally responds slower because of the added circuitry. The response of an ALU is a very important concept in the operation of a digital system.

ALUs are capable of 4-, 8-, and 16-bit arithmetic operations. Some of the new microprocessor systems employ ALUs that are capable of 32-bit arithmetic operations. 4-bit devices are available on single ICs. Eight-, 16-, and 32-bit ALUs are usually part of a microprocessor chip. The ALU is only one part of a rather complex digital system. This part plays an essential role in the operation of the system. It performs all the arithmetic and logic operations. It is connected to other parts of the system through distribution lines called buses. Functionally, the system calls on the ALU to perform specific operations. Input is supplied to the ALU registers and the output is returned to the system over the same bus lines. Operation is dictated by a control signal. This application uses the ALU as a functional component.

The 74LS181 ALU

The 74LS181 is a 4-bit high-speed parallel arithmetic logic unit of the low-power Schottky transistor-transistor logic family. It can perform 32 logic/arithmetic operations. These are divided into 5 arithmetic operations, 16 logic operations, and 8 combined arithmetic/logic operations. This totals 29, with three logic operations repeated. These operations can all be performed in both positive and negative logic. A number of these operations are rarely used. The ALU is, however, considered to be a rather powerful arithmetic device. This chip is housed on a 24-pin dual-in-line package structure. It has the complexity of 75 equivalent gates. A pin configuration and logic diagram of the internal structure are shown in Figure 6-21.

The 74LS181 is a unique chip. Its internal structure is quite complex. In fact, this chip is by far the most complicated that we investigate in this course. The pin designations are quite different from other chips that we have used. These are defined by the following letter designations:

Letters A_3, A_2, A_1, and A_0 are inputs for the binary word A.

Letters B_3, B_2, B_1, and B_0 are inputs for the binary word B.

Letters F_3, F_2, F_1, and F_0 are outputs for the binary word F that is the result of arithmetic/logic operations performed between words A and B.

A carry input, labeled C_n, that is inverted.

A carry output, labeled C_{n+4}, that is inverted.

Letters S_3, S_2, S_1, and S_0 are select inputs that determine the arithmetic/logic operations being performed on words A and B.

A mode-control input, labeled M, that determines the arithmetic/logic operations performed on words A and B. When $M = 0$, arithmetic operations are performed. When $M = 1$, logic operations are performed.

Two energy source inputs identified as + 5 V and ground (GND).

A carry generate output, labeled G.

A carry propagate output, labeled P

A comparator output, designated as $A = B$.

These are defined by the function tables of Figure 6-22. One table is for active high data applications. Active high refers to 1 as + 5 V and 0 as low, or 0 V. This table lists arithmetic operations that are performed without a carry in. An incoming carry adds a 1 to each operation. The alternate table shows how an active low input produces an active low output. Both tables list the functions that are performed on the operands labeled inside of the logic symbol of the device.

THE 74S381 ALU

The 74S381 is a Schottky TTL arithmetic logic unit/function generator that performs eight binary arithmetic operations on two 4-bit words. A functional table and the pin connections for this ALU are shown in Figure 6-23. The eight arithmetic operations are selected by the three function-se-

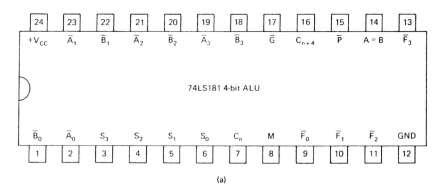

(a)

Figure 6-21. Comparators

Figure 6-21. Comparators (*Continued*)

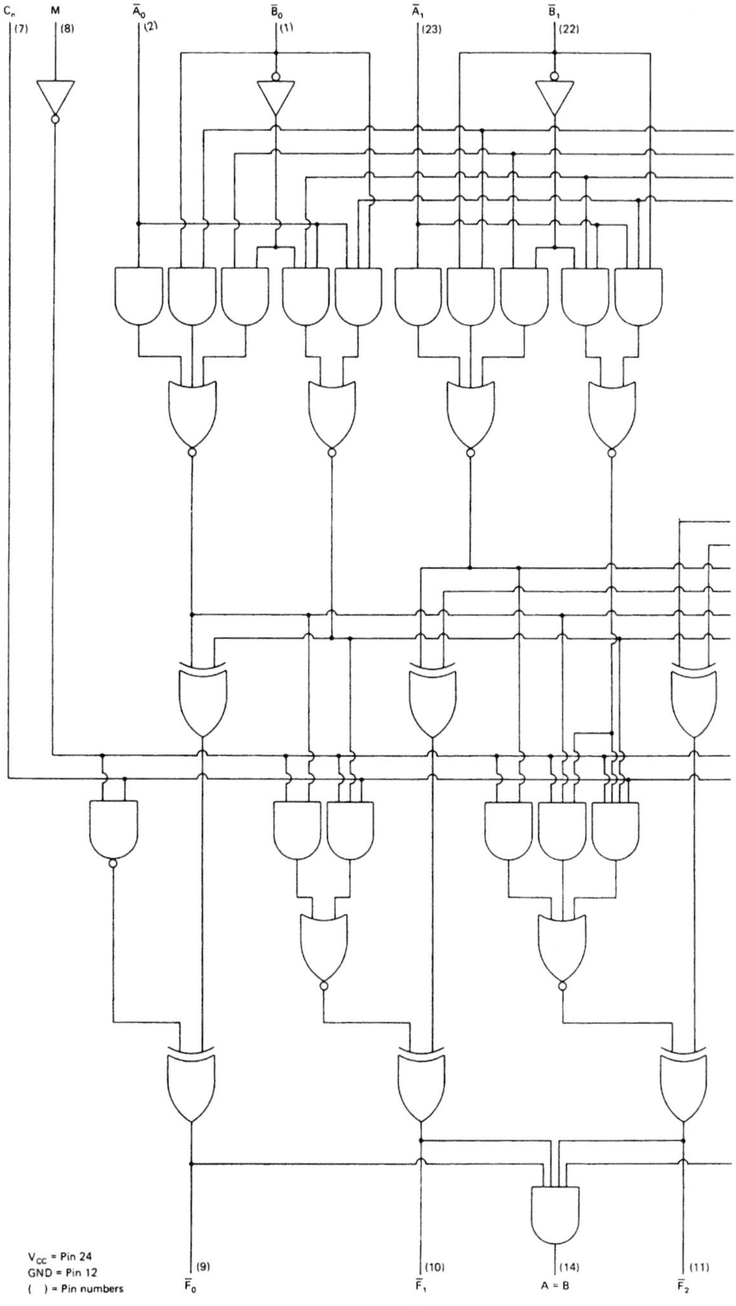

(b)

Figure 6-21. Comparators (*Continued*)

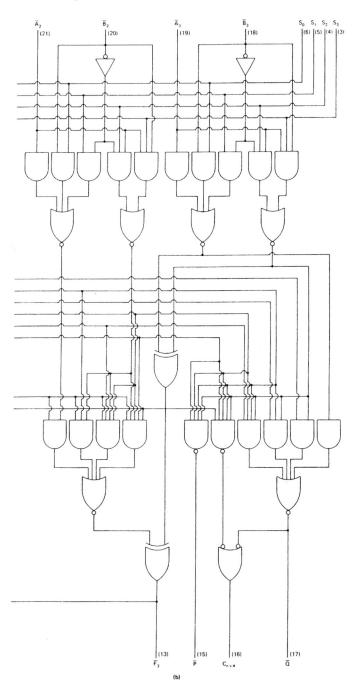

(b)

Mode select – function table

Active high inputs and outputs

S₃	S₂	S₁	S₀	Logic (M = H)	Arithmetic** (M = L)(C$_n$ = H)
L	L	L	L	$\overline{A}$	A
L	L	L	H	$\overline{A+B}$	A + B
L	L	H	L	$\overline{A}B$	A + $\overline{B}$
L	L	H	H	Logical 0	minus 1
L	H	L	L	$\overline{AB}$	A plus A$\overline{B}$
L	H	L	H	$\overline{B}$	(A + B) plus A$\overline{B}$
L	H	H	L	A $\oplus$ B	A minus B minus 1
L	H	H	H	A$\overline{B}$	AB minus 1
H	L	L	L	$\overline{A}$ + B	A plus AB
H	L	L	H	$\overline{A \oplus B}$	A plus B
H	L	H	L	B	(A + $\overline{B}$) plus AB
H	L	H	H	AB	AB minus 1
H	H	L	L	Logical 1	A plus A*
H	H	L	H	A + $\overline{B}$	(A + B) plus A
H	H	H	L	A + B	(A + $\overline{B}$) plus A
H	H	H	H	A	A minus 1

Active low inputs and outputs

S₃	S₂	S₁	S₀	Logic (M = H)	Arithmetic** (M = L)(C$_n$ = L)
L	L	L	L	$\overline{A}$	A minus 1
L	L	L	H	$\overline{AB}$	AB minus 1
L	L	H	L	$\overline{A}$ + B	A$\overline{B}$ minus 1
L	L	H	H	Logical 1	minus 1
L	H	L	L	$\overline{A+B}$	A plus (A – $\overline{B}$)
L	H	L	H	$\overline{B}$	AB plus (A – $\overline{B}$)
L	H	H	L	$\overline{A \oplus B}$	A minus B minus 1
L	H	H	H	A + $\overline{B}$	A – $\overline{B}$
H	L	L	L	$\overline{A}B$	A plus (A – B)
H	L	L	H	A $\oplus$ B	A plus B
H	L	H	L	B	A$\overline{B}$ plus (A – B)
H	L	H	H	A + B	A + B
H	H	L	L	Logical 0	A plus A*
H	H	L	H	A$\overline{B}$	AB plus A
H	H	H	L	AB	A$\overline{B}$ plus A
H	H	H	H	A	A

L = LOW voltage
H = HIGH voltage level
* Each bit is shifted to the next more significant position.
** Arithmetic operations expressed in 2s complement notation.

Figure 6-22. Function tables of a 74LS181 ALU

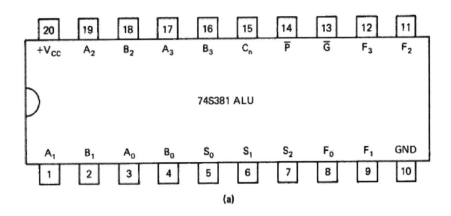

Select inputs			Arithmetic logic operations
S_2	S_1	S_0	
L	L	L	Clear
L	L	H	B minus A
L	H	L	A minus B
L	H	H	A plus B
H	L	L	A ⊕ B
H	L	H	A + B
H	H	L	AB
H	H	H	Preset

(b)

Figure 6-23. Function table and pin configuration of a 74S381 ALU: (a) pin configuration ;(b) function table.

lect lines labeled S_2, S_1, and S_0. The inputs are identified as A_3, A_2, A_1, and A_0 for word A and B_3, B_2, B_1, and B_0 for word B. The outputs are labeled F_3, F_2, F_1, and F_0. C_n identifies the carry input for addition and the inverted carry input for subtraction. P^* is the inverted carry propagate output. G^* is the inverted carry generate output. This chip is housed in a 20-pin dual-in-line package. Its operations are CLEAR, $B - A$, $A - B$, A plus B, A⊕B, A + B, AB, and PRESET.

The 74S381 ALU can perform only eight different arithmetic or logic operations on a pair of 4-bit words. The operations performed are all very useful. Compared with the 74LS181, the 74S381 has limited arithmetic power, but the operations that it performs are widely used. It is ideally suited for high-density system applications. This IC can be cascaded to accommodate larger data words. In general, this chip is easy to use and less complicated than other chips in its circuit applications.

SUMMARY

Addition is a mathematical operation in which two or more numbers or integers are combined together to obtain a simple equivalent quantity. Adding is facilitated by memorizing different integer combinations of the number system.

There are four rules to follow in binary addition: $0 + 0 = 0$, $0 + 1 = 1$, $1 + 0 = 1$, and $1 + 1 = 0$ and a carry of 1. When multiple-bit numbers are added, the same basic rules apply. An exclusive-OR gate achieves binary addition. An X-OR gate and an AND gate form a half-adder. Half-adders are used to add first-order numbers. When two-bit numbers are to be added, a full-adder is needed for the second-order numbers. A full-adder accepts three inputs and generates a sum and carry output.

Subtraction is a mathematical operation in which one integer or number is deducted from another to obtain an equivalent quantity. Subtraction is accomplished by memorizing different number combinations of the system base.

Binary subtraction is performed in the same manner as decimal subtraction. The rules for subtraction of binary numbers are $1 - 1 = 0$, $0 - 1 = 1$ and a borrow, $1 - 0 = 1$, and $0 - 0 = 0$. Binary subtraction can be achieved with an exclusive-OR gate and an AND gate preceded by a NOT gate. The AND/NOT gate is responsible for the borrow function. The X-OR AND/NOT gate forms a half-subtractor. Half-subtractors are used to subtract first-order numbers. Full-subtractors are used to subtract multi-bit numbers. A full-subtractor is formed with two half-subtractors and an OR gate.

Binary subtraction is generally accomplished with adders. The subtrahend is complemented and then added to the minuend. The most significant bit of the difference is removed and added to the remainder to produce the final difference. This is called the 1s complement with an end-around carry. The 2s complement method of subtraction first finds the 1s complement of the subtrahend. A 1 is added to it to produce the 2s complement. The 2s complement subtrahend is added to the minuend to produce the difference. The most-significant place or borrow is dropped. The resulting sum is the difference.

Arithmetic logic units, or ALUs, are used to perform a variety of arithmetic/logic operations on a single IC chip. Some ALUs are stand-alone items, whereas other ALUs are a part of microprocessor chips or computers. ALUs are capable of 4-, 8-, 16-, and 32-bit arithmetic operations. This device responds to binary data applied to its input, and the select inputs are used to determine the arithmetic/logic operations to be performed.

Chapter 7

Digital Timing and Signals

INTRODUCTION

Digital systems can operate with voltage-level changes that occur randomly or can be synchronized by clock signals. Random operation occurs when gates are changed according to the 1-0 level of signals applied to the input. This is generally called asynchronous operation. Synchronous operation is controlled by a clock circuit. A typical clock signal has a series of waves or pulses that occur repeatedly in a given unit of time. This signal is used to trigger a series of operational steps all at the same time. Most digital systems respond to clock signals.

A clock signal is commonly distributed to most of the parts of a digital system. This signal is used to control counting, to distribute data to appropriate locations, and to direct computer functions. In computer systems, clock signals occur at a very high pulse-repetition rate. These signals tell the components of the system when and how to react to the applied data. The clock is often described as the heartbeat of a digital system or computer.

In this unit we are concerned with the generation and shaping of clock signals. An astable multivibrator that switches back and forth between two states is often used as a clock. This circuit can be described as an oscillator. An oscillator changes direct current into a pulsating voltage that occurs at a frequency. Trigger circuits are used to produce clean pulses of energy that will initiate some action in the operation of a system. They can clean up, or reshape, distorted waves, delay pulses, or alter the amplitude so that it conforms to a precise level. Several different trigger circuits can be used to achieve this operation in a digital system.

DIGITAL CLOCK SYSTEMS

A very large portion of all digital operations of a computer are initiated with a clock signal. The clock is a fundamental part of a digital system.

It is responsible for providing a periodic waveform that can be used as a synchronizing signal. The square wave of Figure 7-1 is representative of the output of a clock of a digital system. This signal does not need to be perfectly symmetrical, as indicated by the square wave. It could be a series of spiked pulses or rectangular-shaped waves that occur at a continuous rate. The main consideration is that the waves occur at a periodic rate. In a digital system these waves define the timing interval that controls logic operations. The timing interval is the clock cycle time. It is equal to one period of the clock waveform. Counters, flip-flops, gates, and other logic elements must complete their operation in less than one clock cycle period.

The clock cycle time of a waveform is a very important consideration in the operation of a digital system. This part of a wave represents the point where one complete cycle of operation occurs. It is given as a function of time. A clock signal is generally expressed as a frequency. Frequency is an indication of the number of cycles or hertz (Hz) that occur in a given unit of time. Frequency and time are inversely related. Frequency is equal to the reciprocal of time or $f = 1/T$. Time is equal to the reciprocal of frequency or $T = 1/f$.

It takes a certain time for the wave of a clock to occur. In Figure 7-1, if the frequency is 500 kHz, how long would it take for one wave to occur? Since $T = 1/f$, then the time of one clock cycle time is $1/500 \times 10^3$, or 2 μs. If this clock signal is applied to a JK flip-flop, it must respond in less than 2 μs. As a rule, the operational time of a JK flip-flop is in the range of a 100 ns, or 100×10^{-9} s. Since the clock cycle time is slower than the operational time of the JK flip-flop, it can perform its function when the clock frequency is 500 kHz. Digital system clock frequencies are usually in the range of 1 mHz or more. In some cases this may closely approximate the upper limit of the operational time of a logic device. The clock cycle time of a system must not exceed the propagation time of its logic devices, or it will not function.

The generation of a clock signal for a digital system can be achieved by a number of different circuits. Discrete components such as bipolar transistors, unijunction transistors (UJTs), and programmable unijunction transistors (PUTs) can be used to accomplish this operation. Logic gates can be connected to form these circuits. Specific ICs called precision timers have also been developed for clocks. As a rule, the simplest way to achieve the operation is usually the best. The clock must be capable of generating a usable waveform that has good stability and a predictable frequency. When stability is a factor, some clock circuits are more desirable than others.

Bipolar Transistor Clocks

Discrete bipolar transistors can be connected to form a clock or a time-base generator. Transistors Q1 and Q2 of Figure 7-2 are connected in an astable multivibrator circuit configuration. An astable multivibrator is basically a wave-form generator that is self-starting and operates

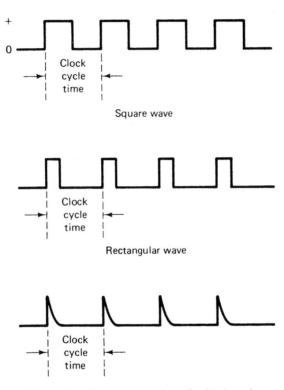

Figure 7-1. Representative clock signals

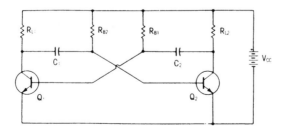

Figure 7-2. Two-transistor astable multivibrator

continuously for long periods of time. The shape of the waveform and its frequency are primarily determined by an RC network applied to each transistor.

Generation of a square or rectangular wave is achieved by connecting two amplifying devices so that the output of each device is connected to the input of the other. The pulse repetition rate (PRR) of this circuit is dependent on the different R and C values of each amplifier. When the component values of each amplifier are the same, the on and off pulse widths are equal. When corresponding sizes differ in value, the pulse width of the output wave has a rectangular shape. Generators of this type are often called multivibrators. Some multivibrators produce a consistent output waveform that is not controlled or triggered into operation by an outside source. The terms astable and free-running are both used to describe this type of multivibrator.

The operation of an astable multivibrator depends on the saturation and cutoff conditions of a transistor. When the source voltage (Vcc) of Figure 7-2 is first applied to the circuit, one transistor starts to conduct before the other. An extremely minute difference in component values causes this to occur. In this circuit, assume that Q_1 goes into conduction first. When this occurs, the collector current (I_c of Q_1 rises quickly, and the voltage across the load resistor RL_1 increases in value. Transistor Q_1 then becomes extremely low resistant, which provides an easy charge path for C_1 through RB_2. The charging current of C_1 passing through RB_2 causes the base of Q_2 to swing negative. As a result, Q_2 is driven into cutoff immediately. At this same time, C_2 begins charging through the emitter-base (E-B) junction of Q_1 and RL_2. As a result of this charging action, the base current of Q_1 is increased, causing it to go into saturation. The process continues according to the time constant of RB_2 and C_1. When this charging current drops off, the base of Q_2 comes out of reverse biasing and Q_2 goes into conduction. This in turn causes Q_2 to become low resistant, which completes a charge path for C_2 through RB_1 and Q_2. Charge current through RB_1 immediately causes it to drive Q_1 to cut off. Capacitor C_1 now begins to charge through the E-B junction of Q_2 and RL_1. This adds to the base current of Q_2, which drives it into saturation. From this point on, the operation cycle is repeated, with Q_1 and Q_2 alternately changing conduction states.

The output of a multivibrator can be taken across either one of the two transistors. The shape of the waveform is primarily based on the time constant of RB_2 x C_1 and RB_1 x C_2. A symmetrical square-wave output occurs when these values are equal. The frequency of a wave generated by this circuit is determined by the expression $f = 1/1.4$ RC. This assumes that $R_1 = R_2$ and $C_1 = C_2$. The duty cycle of this waveform is 50%.

Unijunction Transistor Clock

A unijunction transistor, or UJT, is a three-terminal, single-junction, solid-state device that has unidirectional conductivity. As illustrated in Figure 7-3, a small bar of N-type silicon is mounted on a ceramic base. Leads attached to the silicon bar are called base 1 (B_1) and base 2 (B_2). The emitter *(E)* is formed by fusing an aluminum wire to the opposite side of the silicon bar. The emitter is oriented so that it is closer to B_2 than B_1. A P-N junction is formed by the emitter and the silicon bar. The arrow of the UJT symbol points toward the base which indicates that the emitter is P and the silicon bar is N material.

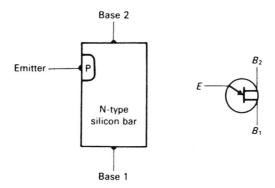

Figure 7-3. UJT crystal structure and schematic symbol

The operation of a UJT is quite different from other solid-state devices. When the emitter is reverse biased, B_1 and B_2 responds as a voltage divider. The resistance of B_1-B_2 is of a maximum value. Any I_E that flows at this time is due to the leakage current of E-B_1. When the emitter is forward biased, RB_1 drops to a very small value. A change in E-B_1 voltage therefore causes a significant change in the resistance of B_1-B_2. The current flow between B_1 and B_2 increases, and V_{BB} decreases in value.

Figure 7-4 shows a UJT relaxation oscillator that is frequently used as a clock. In this circuit, two parallel paths are formed by the components. Resistors B_1, R_2, and C form the charge path, whereas and the UJT form the discharge path. When the switch is turned on, approximately 4 V appears across R_2 and 6 V across the UJT and R_3. This voltage reverse-biases the E-B1 junction by 6 V.

Resistor R_1 and capacitor C_1 receive energy from the source at the same time that the base path does. Capacitor C_1 begins to develop voltage across it at a rate based on the RC values of R, and C_1. When the charge voltage at C_1 reaches 6 V, it overcomes the reverse biasing of the E-B_1 junction. When this occurs, the junction becomes very low resistant and C_1

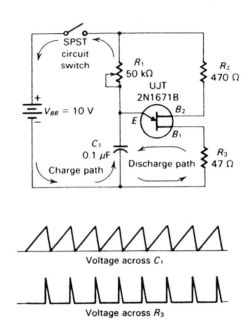

Figure 7-4. UJT clock pulse generator

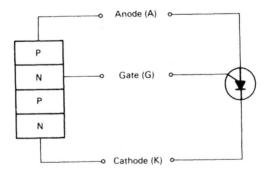

Figure 7-5. PUT crystal structure and symbol

discharges very quickly. The E-B$_1$ junction immediately becomes reverse biased because of the reduced emitter voltage. The capacitor then recharges and the process is repeated. A sawtooth waveform appears across the capacitor, as indicated in Figure 7-4. The discharge voltage across R$_3$ produces a spiked pulse that shows the discharge time of C$_1$. The rise time of the sawtooth wave is adjusted to some extent by different values of R$_1$. When the charge action of the capacitor is maintained in the first time constant, very accurate pulse generation can be achieved by this circuit. The duty cycle of the pulse is a very small percentage of the total operational time.

Programmable Unijunction Transistor Clocks

A programmable unijunction transistor, or PUT, is frequently used as a clock for digital systems. A PUT is actually a three-junction device similar to a silicon-controlled rectifier. It responds as a UJT that has a variable trigger voltage. This voltage can be adjusted to a desired value by changing two external resistors. The voltage needed to trigger the device into conduction can be set or programmed to a specific value.

Figure 7-5 shows the crystal structure, schematic symbol, and element names of a PUT. The crystal structure and schematic symbol of the PUT are very similar to the SCR. The gate junction is the primary difference. In a PUT, the gate *(G)* is connected to the N material nearest to the anode. A P-N junction is formed by the anode-gate. Conduction of the device is controlled by the bias voltage of *A-G*.

The polarity of the bias voltage of a PUT is referenced with respect to the cathode. The cathode is usually connected to the ground or the negative side of the power source. The gate is then made positive relative to the cathode. This is the gate voltage (V_G). The anode voltage (V_A) is also made positive with respect to the cathode. Conduction of the PUT is based on the difference in positive voltage between the gate and anode. When the gate is more positive than the anode, the *A-G* junction is reverse biased. This condition causes the device to be nonconductive, or in its off state. The anode-cathode has infinite resistance and the device responds as an open switch. When the anode becomes more positive than the gate by 0.5 V, it forward biases the *A-G* junction. This condition causes gate current to flow, and the device is triggered into conduction. In the on state, the anode-cathode resistance drops to a very low value. The device then responds as a switch in the on state.

Figure 7-6 shows a PUT relaxation oscillator that can be used for a clock. In this circuit, two parallel paths are formed by the components and the device. Resistor R_3 and C_1 form the charge path for the capacitor and the anode voltage. The PUT and resistor R, form a discharge path for C_1. A trigger pulse is developed across R, when C_1 discharges through the anode-cathode of the PUT.

When power is applied to the circuit of Figure 7-6, it energizes the PUT and the resistor-divider network of R_2 and R_1. The two resistors develop the gate voltage (V_G) as indicated. V_G makes the gate more positive than the anode by a value of approximately 6.6 V. This causes the PUT to be nonconductive. After a short period of time, capacitor C_1 begins to charge to the source voltage through R_3. When the charge potential exceeds 6.66 V, it makes the anode more positive than the gate. This condition triggers the PUT into conduction. C_1 then discharges through the low resistance of the PUT and R_4. Discharge takes place very quickly due to

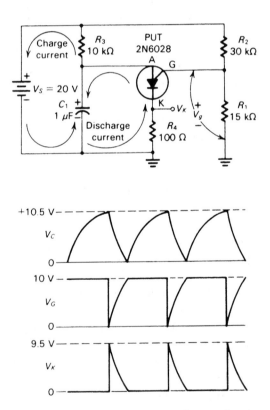

Figure 7-6. Clock circuit and waveforms

the low resistance. The resulting current produces a positive-going voltage pulse across R_4. Triggering the PUT into conduction also causes a drop in the value of VG. This is due to the low resistance of the forward-biased A-G junction. The waveforms of Figure 7-6 show how the relaxation oscillator responds for three operational cycles.

Integrated Circuit Clocks

Discrete component clock circuits have been used for a number of years in digital systems. Clock circuits can be formed equally as well with ICs. IC clocks utilize a minimum of components and have simple circuit construction. The complexity of the circuit is in the internal construction of the IC. Only a few external components are needed to fabricate the circuit. The IC must invert a signal applied to its input. NOT, NAND, and NOR gates can be used to fabricate a clock. A precision timing IC is also available for clock construction.

THE INVERTER CLOCK

The inverter, or NOT, function of an IC can be used to construct a clock. The inverting function is used to achieve the same operation as the two transistors of the discrete component astable multivibrator of Figure 7-2. The output of each transistor is coupled to the input of the other transistor. This method of coupling caused the circuit to have regenerative feedback. This helps the circuit to overcome internal resistance and produce continuous oscillation. Two inverters can be used to accomplish this same operation. An external resistor-capacitor network is added to the inverter circuit to produce timing.

An inverter clock is shown in Figure 7-7. This clock has two outputs labeled Q and $\bar{Q}$. These can also be called clock phase 1 (for Q) and clock phase 2 (for $\bar{Q}$). When the output of Q is high, $\bar{Q}$ will be low and vice versa. These outputs will never be at the same level simultaneously.

A rather small-valued resistor (R_1) is used for the resistor of the time constant. This resistor is primarily used to bias the inverter stage near the linear part of its operating characteristic. Typical values are in the range of 150 to 270 Ω Capacitor C_1 is used to feedback the signal from inverter 2 to inverter 1. C_1 is the primary factor in determining the frequency of the clock. The value of C_1 may be changed to establish the correct timing sequence.

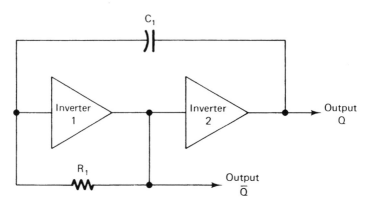

Figure 7-7. An inverter clock

The period *(p)* or time *(T)* of one clock cycle is $3R_1C_1$. Since frequency *(f)* = 1/T, then $f = 1/3(RC)$.

Operation of the inverter clock is based on the operational state of each inverter. The output of inverter 1 is connected directly to the input of inverter 2. The output of inverter 2 is connected to inverter 1 through C_1. Resistor R_1 is connected across inverter 1. Inverters connected in this manner will have regenerative, or in-phase, feedback between the circuit's input and output. If the input to inverter 1 is high, its output will be low. This is coupled to the input of inverter 2. Its output is high and returned to the input of inverter 1. The feedback signal from 2 to 1 is in phase with the original input signal applied to inverter 1. The timing operation of the circuit is controlled by the value of R_1 and C_1.

Let us now look at the operation of the inverter clock in a sequence of events. These steps are shown in Figure 7-8 and summarized as follows:

1. Assume now that output of inverter 2 goes low as a result of some difference in the component values of the IC. This low is immediately coupled to the input of inverter 1 by capacitor C_1. The output of inverter 1 goes high, which causes a change in the input of inverter 2. This causes the output of inverter 2 to remain in its beginning low state.

2. At this point in time, capacitor C_1 begins to charge through R_1 toward the high state at the output of inverter 1. Note the direction of the charging current flow in the drawing for step 2.

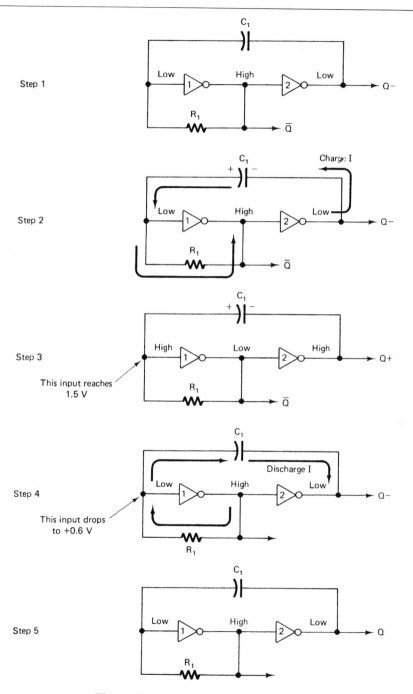

Figure 7-8. Inverter clock operation

3. C_1 continues to charge at the rate of R_1C_1. Since R_1 is very small, the charging action occurs very quickly. When the voltage at the input of inverter 1 reaches approximately + 1.5 V, it causes a state change. This changes the output of inverter 1 and the input of inverter 2 to a low state. A low input to inverter 2 causes its output to be high. Step 3 shows this condition. The high output of inverter 2 and the charge on C_1 causes the input of inverter 1 to remain high.

4. Capacitor C_1 begins to discharge. When the input voltage of inverter 1 drops to something less than + 0.6, it is considered to be in its low state. This causes the output of inverter 1 to go high. This in turn causes the output of inverter 2 to go low. Step 4 shows this condition of operation.

5. One period of operation has been completed by this time. Step 5 shows the output of inverter 2 to be low, which was the starting point of the operational cycle. From this point on the operation is repeated.

With the output of each inverter connected to the input of the other inverter, it is possible to generate a continuous clock signal. The shape of the wave depends on the values of resistor R_1 and capacitor C_1. In an actual circuit, the leading and trailing edges of the wave may be rounded for a long time count and spiked for a short time count. This is usually considered to be a trade-off with respect to frequency selection. This type of clock produces a wide range of frequency possibilities but does not produce a good square wave. To solve this problem, the Q output is often connected to another inverter. This isolates the time constant components of the clock from its output and helps to shape the wave. If a two-phase output is desired, both Q and $\bar{Q}$ outputs can be connected to an independent inverter. Since most ICs have six inverters in their makeup, this is not a significant problem in circuit construction.

NAND and NOR Gate Clocks

NAND and NOR gates can be used to construct a clock by using the gates as an inverter. The circuit is primarily the same as the inverter circuit of Figure 7-8. An inverter can be accomplished by connecting the inputs of a NAND or NOR gate together. Feedback is achieved by connecting a capacitor from the output of one gate to the input of the other gate. A resistor is coupled between the two inputs. The period *(p)* of operation is $3R_1C_1$ and the frequency *(f)* is $1/3R_1C_1$.

NAND and NOR gates distort the generated output wave to some extent. To reduce this problem, the Q and Q′ outputs of the clock are connected to an *RS* flip-flop. The flip-flop responds as a bounceless switch. The distorted output of Q and Q′ is reduced with this addition. This accomplishes the same thing as connecting inverters to the output of Figure 7-7. Figure 7-9 shows clocks built with NAND and NOR gates. The output is a rather good square wave for low and medium frequencies.

Crystal-controlled Clocks

The inverter clocks of Figs. 7-7 and 7-9 are not stable at frequencies above 100 kHz. The *RC* components of the circuit are responsible for the instability. When a stabilized output signal is required for system operation, a crystal-controlled clock must be used. A crystal may be added to the inverter clock to produce stability.

Figure 7-10 shows a crystal-controlled inverter clock. In this circuit the crystal is connected between the output of inverter 1 and the input of

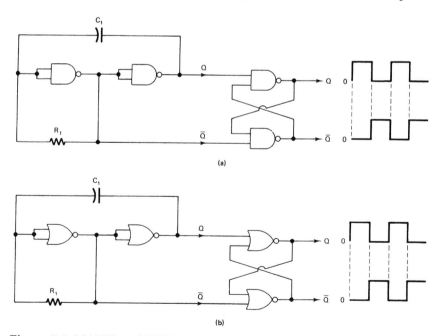

Figure 7-9. NAND and NOR gate clocks: (a) NAND gate clock; (b) NOR gate clock

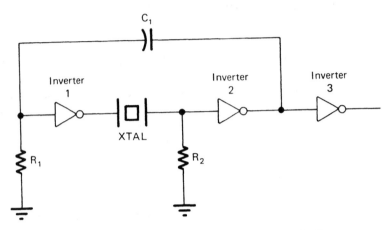

Figure 7-10. Crystal-controlled inverter clock

inverter 2. Resistors R_1 and R_2 are used to bias the inputs of the inverters at approximately 1.5 V. This causes each inverter to operate in the linear region. Capacitor C_1 isolates the bias voltage of inverter 1 from the output of inverter 2. Capacitor C_1 and resistors R_1 and R_2 are selected to cause oscillation near the natural frequency of the crystal. When the frequency of the circuit changes at a rate near the natural vibrating frequency of the crystal, the circuit will eventually lock onto the crystal frequency. The crystal then maintains the frequency at a constant value. Crystal-controlled oscillators are used when it is desirable to have accurate clock frequencies that are stable for long periods of time.

IC CLOCKS

Special IC chips are now available that have the capability of producing a variety of timing functions. This development has opened the way for a number of new and unusual approaches to electronic timing. Through this device, timing operations can now be achieved with a minimum of components. One very important application deals with the generation of digital time-base waves. The precision timer of the 555 series was one of the first developments in this area. It was first introduced by Signetics Corporation. Several variations of the basic timer are now available through other manufacturers. Figure 7-11 shows a functional block diagram of the timer chip.

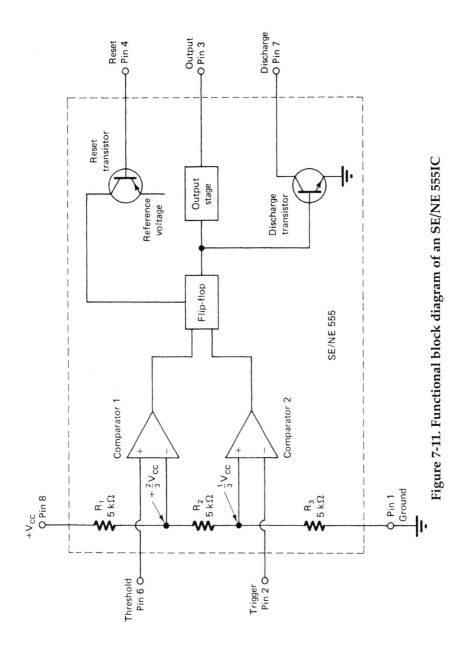

Figure 7-11. Functional block diagram of an SE/NE 555IC

The operation of an SE/NE 555 timer is directly dependent on its internal functions. The three resistors, for example, serve as an internal voltage divider for the source voltage. One-third of the source appears across each resistor. Connections made at the 1/3 and 2/3 Vcc points serve as reference voltages for the two comparators. A comparator is an op-amp that changes states when one of its inputs exceeds the reference voltage. Comparator 2 is referenced at +1/3 Vcc. If a trigger voltage applied to the negative input of this comparator drops below +1/3 Vcc it causes a state change. Comparator 1 is referenced at +2/3Vcc. If voltage at the threshold terminal exceeds this reference voltage, the comparator goes through a state change. Note that the output of each comparator is connected to an input of the flip-flop.

The flip-flop of the SE/NE 555 timer is a bistable multivibrator. It changes states according to the voltage value of its input. If the voltage value of the threshold terminal rises in excess of +2/3 Vcc it causes the flip-flop to change its state. A decrease in trigger voltage below +1/3 Vcc also causes comparator 2 to change its state. This means that the output of the flip-flop is controlled by the voltage values of the two comparators. A state change occurs when the trigger input drops below +1/3 Vcc or when the threshold rises above +2/3 Vcc.

Note that the flip-flop output of the SE/NE 555 is used to drive the discharge transistor and the output stage. A high, or positive, flip-flop output is used to turn on both the discharge transistor and the output stage. This condition causes the discharge transistor to be conductive and causes it to respond as a low-resistance short to ground. The output stage responds in the same manner. A reversal of this action takes place when the flip-flop output changes to a low, or 0, state. The discharge transistor then responds as an open circuit or becomes infinite. The output state swings to its high, or positive, Vcc state. In effect, the operational state of the output and discharge transistor is based on voltage applied to the trigger or threshold input terminals.

Figure 7-12 shows an SE/NE 555 IC timer connected as an astable multivibrator. In this time-base generator, pins 8 and 4 are connected to the dc energy source of somewhere between 5 to 15 V dc. When pins 2 and 6 are connected, the 555 triggers itself and operates as a free-running multivibrator. The external capacitor (C) charges through both resistors R_A and R_B and discharges through R_B and the internal parts of the IC. The duty cycle, which is total cycle time divided by the on time, is precisely set by the ratio of these two resistors.

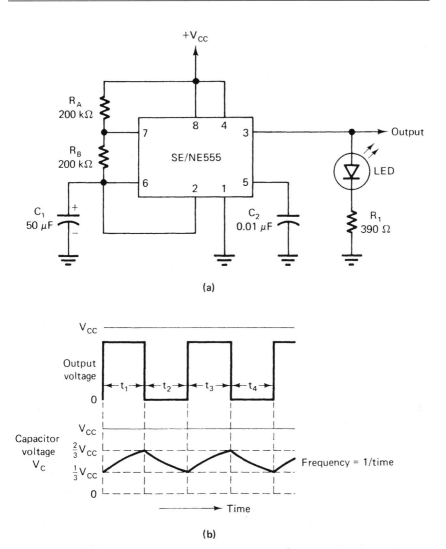

Figure 7-12. Astable multivibrator using an SE/NE 555IC: (a) circuit; (b) waveforms

In actual circuit operation, the capacitor charges from $1/3$ to $2/3$ of the Vcc voltage. It then discharges through R_B and the IC to a value of $1/3$ Vcc. As shown by the waveform graph of Figure 7-12b, as C_1 starts to charge, the output goes to a high value. It remains at this level for the time period t_1. In seconds, this represents 0.693 of $(R_A + R_B)C$.

When C_1 begins to discharge through RB, the voltage drops to one-third of the value of Vcc. This is represented by the time that the output drops to a low value, or t_2. In seconds, t_2 is $(0.693\ RB)C$.

The output frequency of an astable multivibrator is represented by the total time required to charge and discharge C. The combined period of $t_1 + t_2$, $t_2 + t_3$, or $t_3 + t_4$ is represented by a capital letter such as T. The frequency of a total operational cycle is therefore a function of time. In practice, the formula $f = 1/T$ shows this relationship. For a multivibrator, the following formula is used:

$$\text{frequency in hertz} = 1.44\ /\ [\ (R_A + R_B)C\ ]$$

The resistance ratio of R_A and R_B of an IC clock is quite critical. If R_B is more than half the value of RA, the circuit will not operate. Essentially, this would not permit pin 2 to drop in value from 2/3 Vcc, to 1/3 Vcc This would not allow the IC to retrigger itself, which prepares it for the next time period. IC manufacturers usually supply design data information of this type that can be used to select proper resistor-capacitor ratios for a desired operating frequency.

MONOSTABLE MULTIVIBRATORS

A monostable multivibrator is used to generate time intervals with 0.1 to 10% accuracy. It is also called a single-shot, or one-shot, multivibrator. This multivibrator has a stable state and an unstable state of operation. It must be triggered to change states. Triggering causes the multivibrator to change from a stable state to an unstable state and then return to a stable state. An output pulse delivered by this circuit has a fixed duration. The duration of the pulse is usually determined by passive components such as a resistor and a capacitor. The duration may be fixed or variable, depending on component selection. By connecting several of these circuits together, a variety of sequential logic operations can be achieved. Monostable multivibrators can also be used to reform trigger pulses, and to reshape waves.

Discrete Component Monostable Multivibrators

Figure 7-13 shows the circuitry of a discrete component mono-stable multivibrator. This circuit employs bipolar transistors in its construction.

The circuit is very similar to the astable multivibrator of Figure 7-2. Only one capacitor is used in the one-shot circuit. This removes the continuous charge and discharge action of the circuit. As a result, it is not self-starting. The one-shot circuit will be in a stable state when it is energized. When a trigger pulse is applied to the input, it will change from stable to unstable and return to its original stable state. The duration of the transition from stable to unstable is determined by the value of C_1 and R_1.

Assume now that the circuit of Figure 7-13 is energized by the V_{CC} source. Initially, Q_2 is forward biased by R_3 and R_5. This immediately causes Q_2 to saturate. The collector voltage of Q_2 drops to a very small value. This is coupled to the base of Q_1 through resistor R_2. Q_1 has insufficient bias voltage to be conductive. It is driven to cutoff. Capacitor C_1 charges through R_1 and the base-emitter junction of Q_2. The capacitor charges very quickly through these components. The circuit is in its stable condition of operation. The output of Q_2 is low for this state. The circuit will remain in this state as long as energy is supplied by the source.

To initiate a state change in a monostable multivibrator, a trigger pulse must be momentarily applied to the input. A short-duration square wave applied to C_2 and R_5 will produce a differentiated positive and negative spike across R_5. See this at the circuit input. A diode (D_1) connected to the base of Q_1 will be conductive only on the negative part of the spiked wave. This causes a negative spike to be applied to the base of Q_2. Since

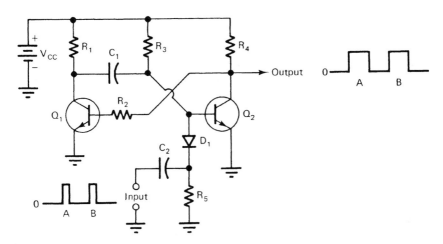

Figure 7-13. Discrete component monostable multivibrator 74123 74LS123

Q_2 is an NPN transistor a negative pulse of sufficient value will reverse-bias the base. This will immediately drive Q_2 into cutoff. With Q_2 in a nonconductive state, its collector voltage rises to the value of V_{CC}. This is immediately applied to the base of Q_1, causing it to saturate. C_1 then discharges through R_3 and the collector-emitter of Q_1. The discharge current through R_3 causes the base of Q_2 to continue to be negative and remain cut off. After a short period of time, the discharge current of C_1 drops to a very small value. This time is determined by 0.693 R_3 x C_1. When the discharge current stops, it immediately causes the base of Q_2 to be forward biased. This causes Q_2 to be conductive and the collector voltage to drop to a very low value. This represents the low, or stable, operating state. The output is low and the circuit has cycled from stable to unstable and back to a stable state. The duration of the transition is dependent on the value of C_1 and R_3. For a long transition period, either value can be increased. This permits the shape of a trigger pulse to lengthened or reformed according to the demands of the system.

IC Monostable Multivibrators

Most of the monostable multivibrators in use today are of the IC type. Operation of the IC is primarily the same as the discrete component circuit. Most IC families have two or three versions of the monostable multivibrator. The TTL family of one-shots is investigated in this unit. The monostable multivibrators of other IC families respond in the same manner and have similar characteristics.

An IC monostable multivibrator generally has some additional components other than the flip-flop. The 74123 one-shot of Figure 7-14 has two monostable multivibrators in a dual-in-line package. Note the symbol of each one-shot. It has a flip-flop driven by an AND gate, and a NOT gate is attached to one of the AND gate inputs. The outputs are Q and $\bar{Q}$. When energized, Q is normally low and $\bar{Q}$ is high. This is the stable operational state. When a trigger pulse arrives at the input the output changes with Q going high and $\bar{Q}$ going low. After a time delay, Q and $\bar{Q}$ revert to their normal states. The delay time is controlled by an external connected resistor and capacitor network. Each one-shot is completely independent of the other. They are supplied operational energy through common power supply connections.

The gate circuitry on the input of the flip-flop is used to select the triggering polarity. There are two ways to trigger the flip-flop. If input A is held low, bringing input B from low to high produces triggering. If in-

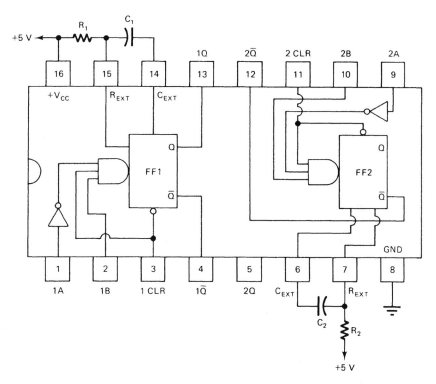

Figure 7-14. Duel retriggerable monostable multivibrator 74123 74LS123

put B is held high, bringing input A from low to high produces triggering. This permits the monostable multivibrator to have selective pulse triggering. The flip-flop also has a clear input connection. In normal operation the clear input (CLR) should be high. Being grounded or connected to a low level inhibits triggering and returns the circuit to its stable state with Q low and $\bar{Q}$ high. This operation permits the flip-flop to change back to its stable state even after it has been triggered and is using its delay function. Clearing simply resets the operation cycle, where it will remain until the next trigger pulse arrives.

The 74123 is classified as a retriggerable monostable multivibrator. Typically, a one-shot multivibrator necessitates some time to recover from a trigger pulse. Once the circuit has been triggered and times out, it takes a certain period of time for the capacitor of the RC network to be recharged through the circuit resistance. This recovery time limits the upper duty cycle to approximately 90%. A retriggerable monostable circuit eliminates this problem. The recovery time of the capacitor is practically instantaneous. This makes it possible for a 100% duty cycle to be achieved. A 100% duty cycle will provide a constant 1, or high, output at Q.

A retriggerable monostable multivibrator permits the generation of very long duration output pulses. When the external R and C values produce an output duration that is longer than the interval between input trigger pulses, the output remains in its unstable state for a long period of time. Figure 7-15 shows waveforms that illustrate this condition of operation. Assume now that triggering occurs on the trailing edge of input pulse 1. This causes output Q to go high, or 1. The monostable multivibrator is in its unstable state of operation. R and C determine the time duration of the unstable state. If RC is longer than the time between pulse intervals, the output will not change. When the trailing edge of pulse 2 arrives, it will automatically terminate the first interval and start a new interval. This all takes place so quickly that Q continues at its high state. With the arrival of pulse 3, the process is repeated. After pulse 3, there is a long period void of pulses. The output will eventually time itself out when a trigger pulse does not appear at the input. Retriggering causes the output to be continuously updated. This causes the output to conform with a long series of short-duration input pulses and produce a single long-duration output pulse.

The external timing function of a 74123 monostable multi-vibrator is accomplished by connecting a resistor and a capacitor to pins 14 and 15 or pins 6 and 7. The pulse duration is $P_d = 0.25RC (1 + 0.7/R)$. If the resistor is 1 kΩ or more, the bracketed part of the formula can be omitted. For this circuit application, the formula is simply $P_d = 0.25RC$. If the value of R is

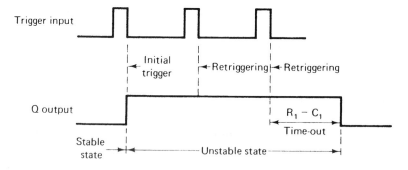

Figure 7-15. Retriggering function of a monostable multivibrator

less than 1 k11, the bracketed part of the formula should be included in the expression. The *RC* components are independent of the IC and can be altered to meet the demands of the application. As a rule, *RC* selection is made to have a long time-out for most applications.

555 Timer Monostable Multivibrators

The precision 555 timer can be used as a monostable multivibrator. This circuit will generate a single output pulse in response to a trigger pulse applied to its input. The duration of the output pulse depends on the values of external components C and R_A. Figure 7-16 shows the 555 connected in a monostable circuit configuration. Note that only one resistor (R_A) and one capacitor (C) are needed in a monostable circuit. The internal structure of the 555 is shown to demonstrate the operation of the device.

A monostable multivibrator is designed primarily to generate a single output pulse in response to an input trigger pulse. The duration of the output pulse depends entirely on the time constant of C and R_A. When a negative-going pulse is applied to the trigger input (pin 2), it causes the output (pin 3) to go high or to + *Vcc*. The trigger pulse causes comparator 2 to drop below its referenced value of + 3 *Vcc*. This action causes the flip-flop to go to its low state. A negative voltage to the discharge transistor causes it to become infinite. This removes the short to ground for capacitor C. The voltage across C begins to rise in value according to the time constant of *RA* and C. When the voltage of C exceeds + 2/3 *Vcc*, causes comparator 1 to change states. This in turn causes the discharge transistor again to become conductive. C then discharges very quickly to ground through pin 7. The output stage follows this action and drops to its low, or ground, state. In effect, the output stage follows the change in the trigger input level. Figure 7-16(b) compares the trigger input and output waveforms.

The duration of the trigger pulse of a multivibrator can be either longer or shorter that the generated output pulse. The resulting width of the output pulse is based on the values of R_A and C. The time *(T)* of the output is based on the formula

$$\text{time (high)} = 1.1 \; R_A \times C$$

This means that the time duration of the output can be altered according to the application. A short-duration input pulse can therefore be used to produce a long-duration output pulse. Because of this, monostable multivibrators of this type are often called *pulse stretchers*.

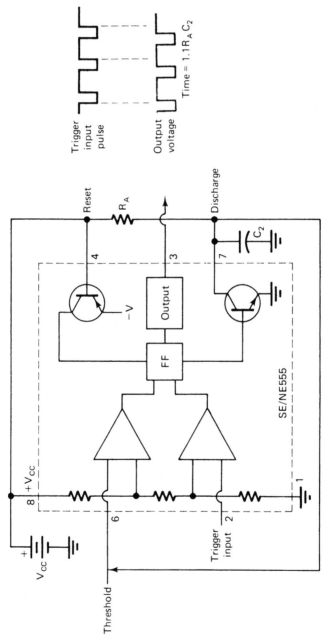

Figure 7-16. A 555 monostable multivibrator

Monostable Multivibrator Applications

The monostable multivibrator has a number of unique applications in digital systems. The IC family usually dictates these applications. TTL monostable multivibrators are used for pulse stretching, pulse shaping, pulse delay, and pulse sequencing. The precision timer IC is used to delay pulses, stretch pulses, and as interval or delay timers. CMOS ICs are used for all these plus some unusual touch switch applications. The delay function, retriggerable capability, and resetting feature make the monostable multivibrator a versatile electronic component.

Figure 7-17 shows a pulse-stretching application of a monostable multivibrator (MMV). In this application a pulse applied to the input is stretched to a rather long duration. The duration of the output pulse is determined by the *RC* time constant. This value can be altered outside of the chip. Note that the input trigger pulse has a very short duration. When this pulse is applied to the input, it causes triggering on the trailing edge of the pulse.

The RC values of the circuit determines the duration of the output pulse. The actual time out of this circuit is 0.693RC. If this value is longer than the duration of the input pulse, the duration of the output will be stretched, or extended. Pulse stretching or shaping is a primary function of the monostable multivibrator.

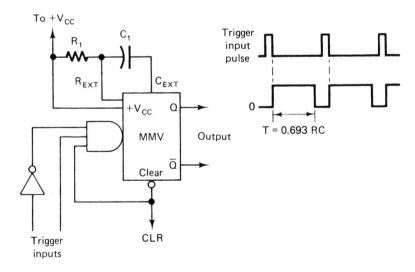

Figure 7-17. Pulse stretching

Figure 7-18 shows a pulse delay application of the monostable multivibrator. In this application two one-shots are used to accomplish the
operation. The Q output of MMV 1 is connected to the input of MMV
2. When the first multivibrator receives a trigger pulse, it causes a state
change. This circuit is set up to trigger on the trailing edge of the trigger pulse. The duration time for this operation is 0.693RC. If a long pulse
delay is desired, the first MMV is responsible for the time-out. When the
first MMV times-out, it triggers MMV 2. This triggering also occurs on the
trailing edge of the wave. Monostable 2 triggers and produces a pulse. The

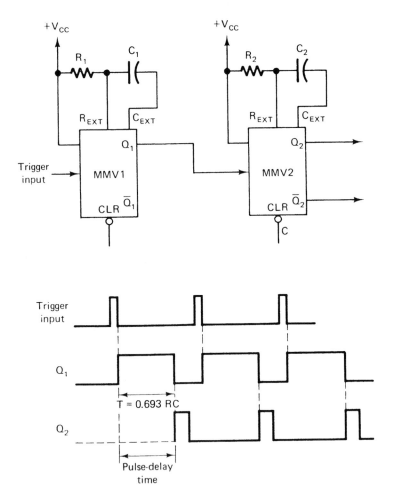

Figure 7-18. Trigger pulse delay with monostable multivibrators

duration of this pulse is dependent on 0.693RC of MMV 2. If the width of the pulse must be the same as the input, the *RC* can be altered accordingly. The time-out of the delay action is dictated by the *RC* value of MMV 1.

Figure 7-19 shows a monostable multivibrator pulse sequence generator. In this application of the one-shot, three MMVs are used. A trigger pulse applied to MMV 1 sets up the sequence. In this case, one pulse will generate three pulses in a sequence. The duration of each pulse is determined by 0.693RC of each MMV. If the pulse width needs to be the same all *RC* values will be equal. If variable pulses widths are required, the *RC* value of each MMV can be adjusted to produce a desired width. This monostable multivibrator generator is used to control a number of

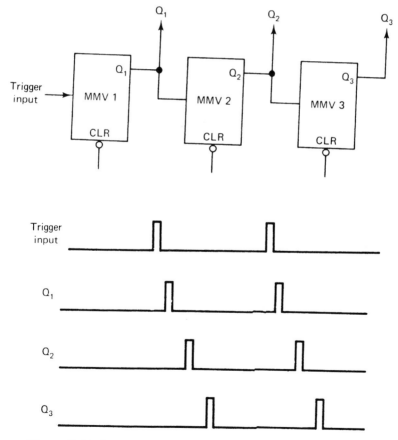

Figure 7-19. A monostable multivibrator pulse sequencer

events in some sequential order. The initial pulse applied to MMV 1 establishes the operational sequence. The pattern of the sequence can be altered by the order in which individual MMVs are connected. The number of events in the sequence is dictated by the number of MMVs used in the generator circuit. The generated event pulse is developed by the Q output of each MMV.

SUMMARY

Synchronous operation of a digital system is controlled by a clock circuit. The clock is responsible for generating a periodic waveform. The clock signal does not need to be perfectly symmetrical. It can be a series of spiked pulses, rectangular waves, or square waves that occur at a continuous rate. These waves define the timing interval that controls logic operations. The timing interval is the clock cycle time. It is equal to one period of the clock waveform Discrete components can be connected to form a clock. Two bipolar transistors are connected to form an astable multivibrator. In this circuit the transistors are connected so that the output of each device is connected to the input of the other. The pulse repetition rate of the circuit is dependent on the RC values of each amplifier. A unijunction transistor, or UJT, can also be used to generate clock pulses. The UJT is used as a relaxation oscillator. This circuit generates a spiked pulse of a short duration. Programmable unijunction transistors, or PUTs, respond in a similar manner. The PUT has a variable value of trigger voltage.

An IC must invert a signal applied to its input in order to form a clock. NAND, NOR, and NOT gates can be used for this operation. An inverter clock has the output of each inverter coupled to the input of the other inverter. The period (p) of a clock pulse is 3RC. The frequency is 1/3RC. NAND and NOR gates are connected as inverters and used to form clocks. Inverter clocks are not as stable at frequencies above 100 kHz. A crystal may be added to an inverter to improve its stability.

The 555 precision timer is a special IC commonly used for clocks. Operation is based on voltage values changing from 1/3 to 2/3 of V_{CC}. This voltage change triggers a comparator, which in turn changes the flip-flop. The flip-flop controls the output and the discharge transistor. In one time period C charges through resistors RA and R_B. In seconds, this is 0.693($R_A + R_B$)C. C discharges through resistor R_B. In seconds, this is $0.693R_B \times C$. The combined period of an operational cycle is $t_1 + t_2$. In frequency this is $f = 1.44/[(R_A + 2R_B)C]$.

Monostable multivibrators, or one-shots, are used to generate time intervals. Triggering causes a one-shot to change from a stable state to an unstable state and then return to a stable state. An output pulse generated by this circuit has a duration that is determined by an external resistor-capacitor network. For a bipolar transistor one-shot, the duration is $0.693RC$. IC monostables are widely used today. These MMVs have retriggering capabilities and trigger polarity selection. Retriggering permits the generation of very long duration output pulses. A 555 can be used as a monostable multivibrator. The time of an output pulse is 1.1 $R_A \times C$. Monostable multivibrators are used for pulse stretching, shaping, delay, and sequencing.

Chapter 8

Sequential Logic Gates

INTRODUCTION

Digital systems employ a number of devices that are not classified specifically as logic gates. These devices play a unique role in the operation of a digital system. Such things as flip-flops, counters, registers, decoders, and memory devices are included in this classification. We discuss truth tables, logic symbols, and the operational characteristics of these devices so that they may be used more effectively when the need arises. As a general rule, most of these devices are constructed entirely on IC chips. Operation is based to a large extent on the internal circuit construction of the IC. Very little can be done to alter the operation of these devices other than to modify the input or use its output to influence the operation of a secondary device.

Digital logic circuits are classified into two groups. We have worked with logic circuits that make up one part of this classification. AND, OR, NOT, NAND, and NOR gates are considered to be combinational logic circuits. The other part of this classification deals with sequential logic circuits. Sequential circuits involve some form of timing in their operation. The timing function permits one or several devices to be actuated at an appropriate time or in an operational sequence.

Logic gates are the building blocks of combinational logic circuits. A flip-flop is the basic building block of a sequential logic circuit. This unit of the course deals with some basic flip-flop circuits. In a later unit, we look at how the flip-flop is connected to form counters, shift registers, and some memory devices.

THE FLIP-FLOP

Flip-flops are commonly used to generate signals, shape waves, and achieve division. In addition to these operations, a flip-flop may also be used as a memory device. In this capacity, it can be made to hold an output state even when the input is completely removed. It can also be made to change its output when an appropriate input signal occurs.

The flip-flop is a logic device with two or more inputs and two outputs. The outputs are the complements of each other. Figure 8-1 shows the general symbol of a flip-flop. Notice that the outputs are labeled Q and $\bar{Q}$. Any letter designation could be used to identify the outputs. The letter Q is widely used. Q is considered to be normal and $\bar{Q}$ is the inverted output. The Q output can be in either the high (1) or low (0) state. $\bar{Q}$ is always the reverse of Q. A flip-flop has two operational states: $Q = 0$, $\bar{Q} = 1$ and $Q = 1$, $\bar{Q} = 0$.

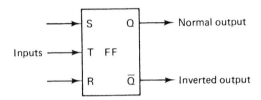

Figure 8-1. Flip-flop symbol

A flip-flop has one or more inputs. These inputs are used to initiate a state change in the operation of a flip-flop. When an input is pulsed or triggered, it will send the output to a given state. It will remain in this state even after the input returns to normal. This is the memory characteristic of a flip-flop.

Flip-flops are known by several names. In general they may be called multivibrators, latches, or by a specific letter designation such as *RS*, *RST*, *D*, or *JK* flip-flop. The letters identify the inputs. *RS* stands for reset and set. *RST* refers to reset, set, and trigger. *D* identifies a flip-flop that has a delay capability. A *JK* flip-flop has data inputs and a clock. Letter-designated flip-flops are more widely used than other identification means.

A Bistable Latch

To explain the operation of a flip-flop, we use the circuitry of a bistable latch. The terms latch and flip-flop can be used interchangeably. The latch used here employs a pair of cross-coupled two-input NOR gates. The term bistable refers to a device that is capable of assuming either one or two stable states or conditions.

A circuit diagram of a bistable latch is shown in Figure 8-2. The truth table of a NOR gate shows its output to be 1, or high, only when the inputs are both 0 or low. A 1 appearing at either or both inputs will produce a 0 output. With this in mind, specific logic states can be assigned to the latch circuit according to the position of the switch. Note the 1-0 designations applied to the inputs of the NOR gates.

Operation of the latch circuit of Figure 8-2 is based on the position setting of the switch. As shown, NOR gate 1 has voltage or a 1 applied to its input through the switch. The input of NOR gate 2 is ground or 0 at the same point in time. A 0 applied to gate 2 causes output $\bar{Q}$ to be 1. This is cross-coupled to the alternate input of gate 1. The second input of gate 1 now being 1 means that Q will be 0. This is cross-coupled to the alternate input of gate 2.

A 0 appearing at both inputs of gate 2 causes $\bar{Q}$ to continue to be 1. The flip-flop will remain in this state as long as power is applied to the circuit. It is latched with Q at 0 and $\bar{Q}$ at 1. Momentarily moving the switch contact off of the upper contact will not change the status of the output.

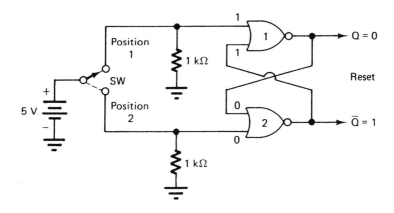

NOR gate truth table

B	A	NOR
0	0	1
0	1	0
1	0	0
1	1	0

Figure 8-2. Bistable latch with NOR gates in the RESET position

The output is latched in the high state. This represents the RESET condition of operation.

To set a latch circuit, the input switch must be placed in the alternate position. Figure 8-3 shows this condition of operation. Note the resulting 1-0 designations for this switch position. Gate 1 has a 0 at one of its inputs and gate 2 has a 1. This causes the Q output of gate 1 to be 1. This is cross-coupled to the alternate input of gate 2. $\bar{Q}$ becomes 0, which is cross-coupled to the alternate input of gate 1. This action latches gate 1 to its 1 state. The gates remain in this latched position as long as power is supplied to the circuit. Momentarily moving the switch off the lower contact will not change the status of the output. A state change can be initiated only by placing the switch in its alternate position. This represents the SET condition of operation.

A pair of two-input NAND gates can also be used to build a bistable latch. The circuit of Figure 8-4 shows NAND gates connected in a latch circuit. The operation of this circuit is similar to that of the NOR latch. This circuit has a modified input. Note that the switch shifts the ground connection to the input of the respective gates. The inverting function of a NAND gate or NOR gate is largely responsible for the operation of a latch.

A bistable latch is widely used in digital electronic circuits. One common application is switch buffering. Mechanical-action electrical switches, for example, have a contact bounce problem when depressed or released. The contacts, which are controlled by a spring action, do not make

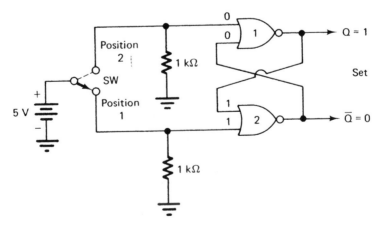

Figure 8-3. Bistable latch with NOR gates in the SET position

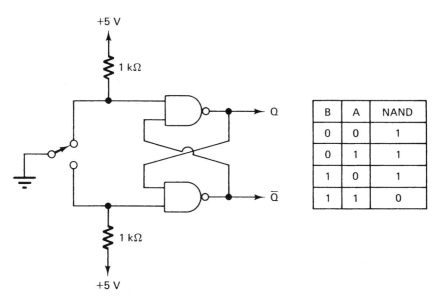

Figure 8-4. Bistable latch with NAND gates

B	A	NAND
0	0	1
0	1	1
1	0	1
1	1	0

an immediate electrical connection or a clean break when actuated. The duration of the bouncing action is generally only a few milliseconds. The contacts actually "bounce" open and closed for a short period of time. This response may cause repeated pulses of energy to appear in a digital circuit. This generally upsets the normal operation of a digital circuit. As a rule, digital circuits must be controlled by bounceless switches.

Figure 8-5 shows an electrical circuit controlled by a mechanical action switch. The single-pole, single-throw (SPST) toggle switch turns the circuit on and off by a shift in the electrical contacts. The voltage across resistor R_1 changes value according to the displayed waveforms. Notice the voltage changes that occur when the switch is actuated. A similar response occurs when the switch is turned off. These voltage transitions are produced by the spring action of the electrical contacts. In digital circuits a switch must be debounced in order for it to be useful.

A latch is commonly used to debounce electrical switches used in digital circuits. The latch circuits of Figs. 8-2, 8-3, or 8-4 could be used as a debounced a switch. Assume now that the latch circuit of Figure 8-3 is used to debounce a switch. When the switch is in position 1, it makes output Q high. This represents the on condition of operation. Changing the switch to position 2 causes the input of the upper NOR gate to go high.

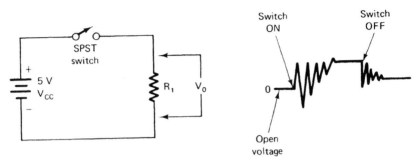

The electrical effect of contact bounce

Figure 8-5. An electrical circuit controlled by a mechanical switch

This causes $\bar{Q}$ to change to a 0, or low, state when it makes the first contact. This change will take place in a few nanoseconds. The gate will stay at this level even if the contacts bounce a few times. Changing the switch back to position 1 will reverse the action. The lower NOR gate will go low, causing Q to go high or turn on. This means that Q changes only when the first pulse of energy is applied. Contact bouncing has no effect on the response of the output once the initial transition occurs. Latches are widely used to debounce digital switching circuits.

The *RS* Flip-flop

An *RS* flip-flop can be constructed from two cross-coupled NAND or NOR gates similar to the bistable latch. Figure 8-6 shows a symbol of the flip-flop, a circuit achieved with cross-coupled NAND gates, a truth table, and waveforms of an operating flip-flop. The outputs of the symbol and NAND gate circuit are labeled Q and $\bar{Q}$. The inputs are labeled R and S. R identifies the reset and S is the set function. The Q and $\bar{Q}$ outputs of the flip-flop are dependent on the voltage level of the *RS* inputs.

The *RS* flip-flop symbol of Figure 8-6 is slightly different than that of *Figure* 8-1. Each input has a small circle attached to it. This is used to denote that the inputs are active low. In the truth table, note that the S input must be low to produce a 1 or high at the Q output. The reset must be 0 to cause $\bar{Q}$ to be 1 or high. The circle at the input therefore identifies the state needed to produce a change. No circle at the input indicates a 1 or high level activated device. A circle at the input indicates a low-, or 0, level activated device. A cross-connected NAND gate is an active low flip-flop and a cross-connected NOR gate circuit is an active-high-level device. The

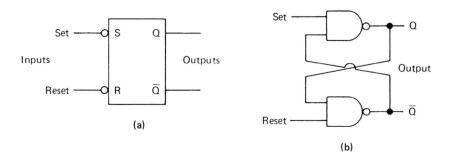

(a)

(b)

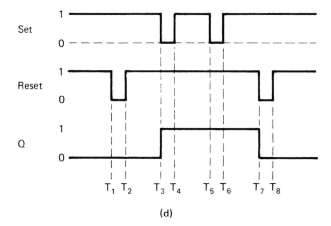

Inputs		Outputs	
S	R	Q	$\overline{Q}$
0	0	1/0	0/1
0	1	1	0
1	0	0	1
1	1	No change	

(c)

(d)

Figure 8-6. Active low *RS* flip-flop: (a) logic symbol; (b) cross-coupled NAND gate circuit; (c) truth table *RS* flip-flop; (d) flip-flop waveforms

presence or absence of a circle on the input of the symbol is a standard method of identifying this characteristic of a flip-flop.

The waveform diagram of Figure 8-6 shows how an *RS* flip-flop responds to changing voltage levels. This is essentially called a timing diagram. It shows voltage level changes with respect to time. The horizontal part of the diagram deals with time and vertical distance is voltage level. This timing diagram is an operational truth table. It shows what you would see on an oscilloscope. Operation of the timing diagram is based on the voltage levels of the *S* and *R* inputs with respect to time. Initially, the SET and RESET inputs are 1 or high. This causes *Q* to remain in its state. When *S* is 1 and *R* is 1 there will be no change in the output. At T_1 the *R* input goes to 0. This does not cause a change in *Q*. When *R* is 0 and *S* is 1, *Q* is 0. Since it was already 0, there is no indicated change in *Q*. At T_2 the reset returns to 1. This does not cause a change in *Q*. At T_3 the SET input changes to low, or 0. This being an active low device causes *Q* to be set to its high, or 1, level. T_4 does not cause a change in *Q* because both S and R are again at the 1 level. At times T_5 and T_6, there will not be a change in *Q* because it is already in the 1 state. The only way to cause a state change in *Q* is to cause the RESET to go low. At time T_7 this shows *Q* to be low. T_8 does not cause a change in *Q* because it returns the input to a 1 level. This will not change the output. This timing diagram shows that the output of a flip-flop remembers its last input and will not change until the opposite input is activated.

The *RS* flip-flop of Figure 8-7 is made from cross-connected NOR gates. The symbol of this flip-flop does not have circles attached to the inputs. This identifies the device as a high-level activated flip-flop. The inputs are opposite to that of the NAND flip-flop of Figure 8-6. Compare the truth tables. The truth table of Figure 8-7 shows that the *S* input must be high or 1 to produce a 1 at the *Q* output. This represents the SET operation. The *R* input must be 1 to cause $\bar{Q}$ to be a 1, or high. This describes the RESET operation. When *S* and *R* are both 0, there is no change in the output. Similarly, 1s at *S* and *R* produce an unpredictable output. The waveform diagram shows how the flip-flop responds to changing voltage levels with respect to time.

Clocked Flip-flops

Flip-flops operate as either asynchronous or synchronous devices. Asynchronous operation occurs when the outputs can change state any time one or more of the inputs change. An *RS* flip-flop is an asynchro-

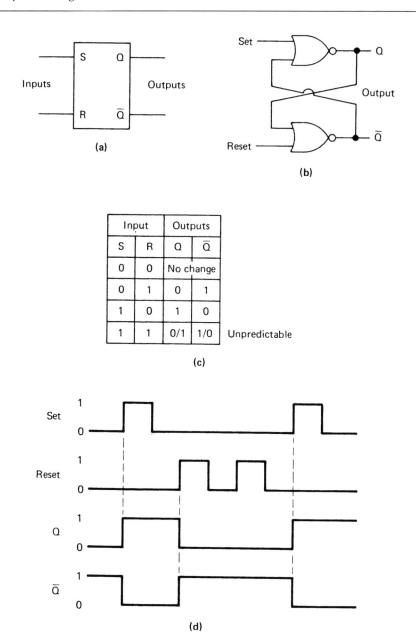

Figure 8-7. High-level activated flip-flop: (a) high-level activated flip-flop; (b) NOR gate *RS* flip-flop; (c) truth table; (d) time waveforms

nous device. Synchronous operation occurs when output changes are controlled by a clock signal. Most flip-flops operate as synchronous devices. Synchronous circuits are easier to troubleshoot because the output can only change at a specific time. Everything is synchronized by the clock signal.

The clock signal applied to a synchronized device is generally a rectangular series of pulses or square waves. The clock signal is distributed to all parts of a digital system. The output of specific devices can make a state change only when the clock signal occurs.

There are two basic types of clock signals used to trigger flip-flops. These are defined as level-triggering or edge-triggering signals. Figure 8-8 shows an example of a level clock signal and two types of edge signals. In level clocking, the state of a clock changes value from 0 to 1 and carries out a transfer of data or completes an action. Data cannot be changed or altered except immediately after a level change. At that time it can be changed only once. In edge clocking, there is positive-edge triggering and negative-edge triggering. Positive-edge triggering occurs at the leading or beginning edge of a pulse. The signal makes a quick transition from 0 to

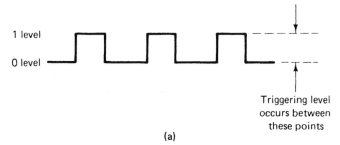

(a)

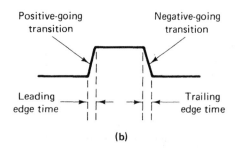

(b)

Figure 8-8. Clock signals: (a) level triggering; (b) wave terms

1. Negative-edge triggering occurs at the trailing edge or end of a pulse. This pulse changes from 1 to 0. An edge-triggered flip-flop can have its input data changed at any time. One square wave has a leading edge and a trailing edge in its make up. A flip-flop can be triggered by only one type of pulse.

Most of the flip-flops used in digital systems are classified as clocked flip-flops. The following are some principle ideas that are common to all clocked flip-flops.

1.   Clocked flip-flops have an input labeled CLK, CK, or CP.

2.   The symbols of clocked *RS* flip-flop are shown in Figure 8-9. Note that the clock input is labeled CLK.

3.   A positive-edge-triggered device is identified by the symbol in part (a). A negative-edge-triggered device is identified by the symbol in part (b). A circle at the clock input indicates negative-edge triggering and no circle indicates positive-edge triggering.

4.   All clocked flip-flops have one or more control inputs. Control inputs are identified by a variety of labels depending on the exact function. The control input determines the state of the output but its effect is not realized until a clock pulse occurs. Essentially, the logic level of an input controls how the output will change, whereas the CLK signal determines when the change takes place.

The Clocked *RS* Flip-flop
    A clocked *RS* flip-flop has reset (R), set (S), and clock (CLK) inputs. The output is labeled Q and $\bar{Q}$. Figure 8-9 shows positive-edge-triggered and negative-edge-triggered clocked *RS* flip-flops. A positive-edge-triggering device changes output states only on the leading, or positive, transition of the clock pulse. It does not respond to negative-edge transitions. Negative-edge triggering takes place on the trailing, or negative, transition of the clock pulse. This device does not respond to positive-edge transitions. The *S* and *R* inputs prepare the flip-flop for a specific state change. The time waveform diagrams show how the output responds to different combinations of the *SR* input. The truth table shows these possibilities. The output of a clocked flip-flop is uniquely different. The outputs are identified as Q_{n+1} and Q. Q_n represents the status of Q before a clock pulse

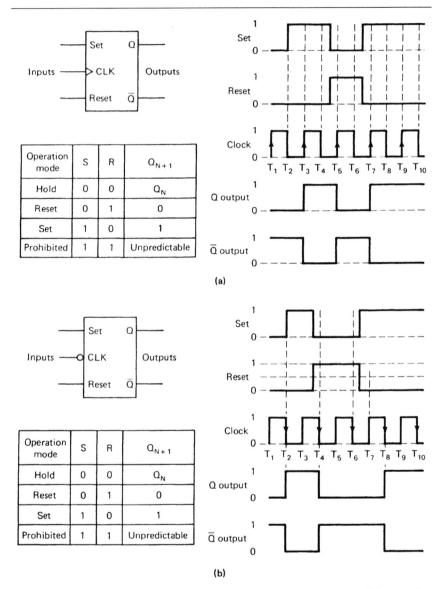

Figure 8-9. Clocked *RS* flip-flops: (a) positive edge triggering; (b) negative edge triggering

is applied. The letter n of this designation refers to the status of Q(now). Q_{n+1} indicates the status of Q after a clock pulse has been applied. The Q_n and Q_{n+1} designations are widely used by IC manufacturers on data sheets.

The operational concept of a clocked *RS* flip-flop is very important. This type of flip-flop is, however, not available on an IC chip. A clocked *RS* flip-flop can be achieved by logic gates. The construction of this flip-flop is very important in the fabrication of other devices. Figure 8-10 shows how a clocked *RS* flip-flop is built with NAND gates. This circuit employs NAND gates 3 and 4 in a cross-coupled *RS* flip-flop. A pulse steering circuit is formed by NAND gates 1 and 2. This part of the circuit is responsible for passing or inhibiting the set and reset input. A clock pulse is needed to enable the steering circuit so that it will pass the input on to the flip-flop. An edge detector circuit is commonly connected to one input of each steering gate. This part of the circuit is used to select the desired edge of the clock signal needed to produce triggering. The combined circuitry of an *RS* flip-flop is often used in the construction of other flip-flops. This circuit provides a convenient method of injecting a clock signal into a flip-flop.

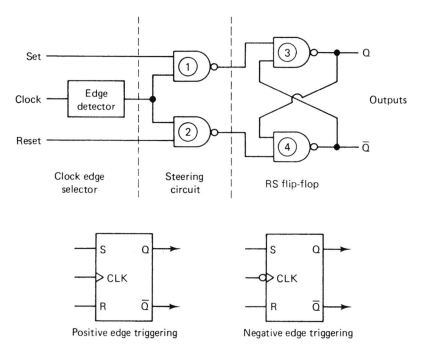

Figure 8-10. A clocked *RS* flip-flop implementation with NAND gates

The Clocked D Flip-flop

An application of the clocked RS flip-flop circuit is the D flip-flop. Figure 8-11 shows the symbol, truth table, and time waveforms of a positive-edge-triggered D flip-flop. The D input, which stands for data or delay, is synchronized with the CLK and determines the state of Q and $\bar{Q}$. Operation is very simple. Q follows the state of D when the positive edge of a clock pulse occurs. This is illustrated by the truth table. D and Q_{n+1}, are the same. Q_{n+1} refers to the output at Q after one (+ 1) positive clock pulse transition. The time waveform shows this action taking place at t_1, t_3, t_5, t_7, t_9, and t_{11}. The data applied to input D has a slight delay action because it does not appear at Q until the positive edge of a clock pulse occurs. The negative edge of the clock pulse has no effect on the transfer of data from D to Q.

Negative-edge-triggered D flip-flops are also available. Operation is

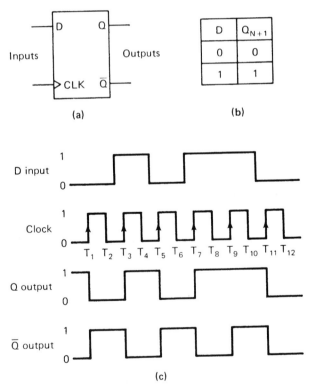

(a)

(b)

(c)

Figure 8-11. D-type flip-flop: (a) positive edge triggering symbol; (b) truth table; (c) time waveforms

the same as the positive-edge triggered devices except triggering occurs only on the negative edge of the clock pulse. The symbol for a negative-edge-triggered flip-flop has a small circle on the clock input. Triggering occurs at t_2, t_4, t_6, t_8, t_{10}, and t_{12} of the time waveform if a negative-triggered device is used.

Figure 8-12 shows how a D-type flip-flop can be constructed with NAND gates. The D flip-flop is simply a clocked RS flip-flop with an inverter connected across its inputs. The inverter causes the inputs to be opposite of each other. This causes the output to follow the value of the D input when it is triggered by the clock pulse. As a rule, several D flip-flops are connected together in circuits to perform data transfer operations. They may be connected in series with Q of one flip-flop connected

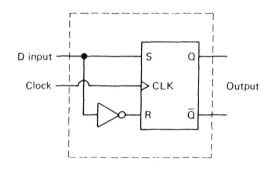

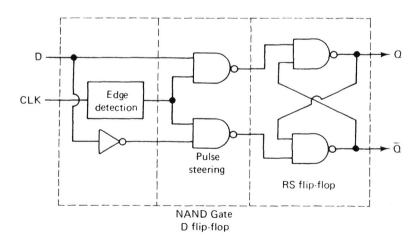

Figure 8-12. D-type flip-flop implementation with NAND gates

to D of the next flip-flop. Parallel-connected D flip-flops transfer data from the input to the output of several devices all at the same time.

The D Latch

An edge-triggered D flip-flop transfers data from the input to the output only when an active clock pulse occurs. If an edge detector is not used on the clock input, the flip-flop will respond differently. A circuit with this type of construction is generally called a D latch.

Figure 8-13 shows a D latch made of NAND gates. Gates 3 and 4 form the cross-connected RS flip-flop. Gates 1 and 2 serve as the steering circuit for the input. An inverter is connected across the two inputs. The clock is commonly applied to the other two inputs. Note that this circuit does not employ an edge detector on the clock input. Compare this circuit with that of the D flip-flop of Figure 8-12.

Operation of a D latch is somewhat different from that of a D flip-flop. A D latch is essentially a level-triggered device. The 0 or 1 level of the clock determines the triggering time. D latches can be triggered from low-speed clock changes or wired DC levels. Operation is described as follows:

1.   When the clock input is low or 0, the D input will not initiate a change in Q. Note this condition in the truth table of Figure 8-13.

When the clock is changed to a high level, the Q output will take on the same level as D. However, if D changes while the clock is high, $\bar{Q}$ will take on the value of D. This means that the triggering operation of a latch is not absolutely dependent on the clock. This condition is described as being *transparent*. It means that clock control is not needed by the flip-flop to achieve all conditions of operation. The edge-triggering detector accounts for this difference in circuit operation.

The logic symbol of Figure 8-13 is slightly different from that of a D flip-flop. The small triangle attached to the clock input of this symbol is omitted. This indicates that the clock is not edge-triggered. It can be triggered by level changes. It should be pointed out that the triangle part of a symbol is not used by some manufacturers. The symbol for a D latch and a D flip-flop may be the same. To avoid confusion in this regard, refer to the manufacturer's data on a specific device instead of the symbol representation. Some manufacturers may use the term enable instead of clock for a D latch. This is done because the clock is supposed to affect the output of

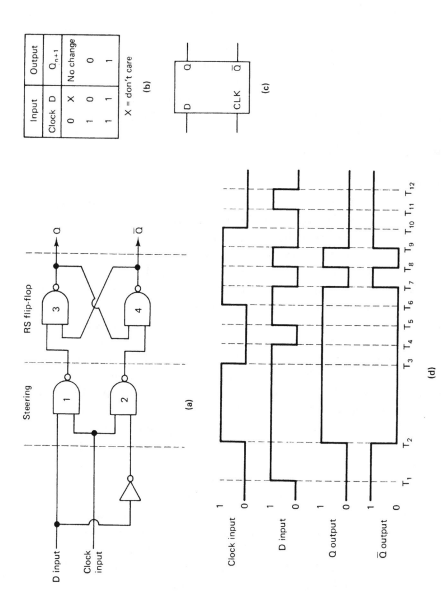

Figure 8-13. *D*-type latch: (a) NAND gate circuit; (b) truth table; (c) *D* latch symbol; (d) waveforms

a device. If it does not affect the output for all conditions, the enable (E or EN) is used. This eliminates the need for a triangle attached to the clock input on some symbols.

A time waveform diagram for a D latch is shown in Figure 8-13. Analysis of the wave diagram shows how the inputs affect the output of a D latch. Note in particular how the output changes at the timed points. At T_1, CLK = 0, D = 1, and Q = 0. T_2 shows CLK, D, and Q all the same. T_3, T_4, and T_5 show when the CLK goes low D does not affect Q. T_6 shows a 1 or high at the CLK, D, and Q. T_7, T_8, and 7_9 show the transparent state of operation. Q changes with D at this time. T_{10}, T_{11}, and T_{12} shows no change in Q when the clock is low. The time waveform displays the operation of the truth table graphically.

The Clocked *JK* Flip-flop

The *JK* flip-flop is probably used more today than all other flip-flops. The *JK* flip-flop is often considered to be a universal device. Its operation is similar to that of other flip-flops. It can, however, be easily modified to achieve different functions. Flip-flops in general have two or possibly three meaningful input combinations. A clocked *JK* flip-flop has four possible input combinations. These are no change, enter 1, enter 0, and toggle. This makes the *JK* flip-flop a very versatile device.

Figure 8-14 shows a symbol and the truth table of a clocked *JK* flip-flop. The *J* and *K* inputs supply data to the device. The letters *J* and *K* do not represent any particular term. They were probably chosen to distinguish this device from the set and reset inputs of an *RS* flip-flop. Control of a *JK* flip-flop is similar in nearly all respects to that of an *RS* flip-flop. When *J* and *K* are both 1, however, this flip-flop does not produce an unpredictable output. It causes the output to change states with each clock pulse. As a result, Q and $\bar{Q}$ both change states continually with the clock. This condition causes the output to respond as a toggle switch. It is generally called the *toggling mode* of operation. Toggling permits a flip-flop to achieve binary division and to shift data one bit at a time.

The *JK* flip-flop of Figure 8-14 is not available today on an IC chip. Flip-flops of this type that are available have some circuit modifications. An SN7476 *JK* flip-flop is shown in Figure 8-15. This device is similar to the basic flip-flop but has two asynchronous inputs called PRESET (PS) and CLEAR (CLR). The *J*, *K*, and CLK inputs are synchronous. The outputs are Q and $\bar{Q}$. The asynchronous inputs are designed to override the synchronous inputs. The truth table shows the response of the asynchronous in-

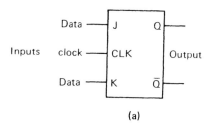

(a)

Mode of operation	Inputs			Outputs	
	Clock	J	K	Q	Q̄
Hold	1	0	0	No change	
Reset	1	0	1	0	1
Set	1	1	0	1	0
Toggle	1	1	1	Toggle	

(b)

Figure 8-14. *JK* flip-flop: (a) symbol; (b) truth table

puts on the first three lines. Since the synchronous inputs are overridden by PS and CLR, an X is used in place of data for these entries. A prohibited state occurs when both PS and CLR are activated at the same time. This occurs when PS = 0 and CLR = 0. This means that the asynchronous inputs are activated by a negative level signal or a dc voltage value. Small circles on the PS and CLR inputs of the symbol denote negative level triggering. It is good practice to avoid using the *JK* flip-flop in its prohibited state. Note that Q and $\bar{Q}$ are both 1 when this mode of operation occurs.

When both asynchronous inputs of a *JK* flip-flop are disabled, the synchronous inputs can again be used to control the output. To disable the asynchronous inputs a 1 must be applied to PS and CLR at the same time. The bottom four lines of the truth table of Figure 8-15 show the status of this operation. Note that the modes of operation are hold, reset, set, and toggle. When PS and CLR are disabled, the flip-flop is controlled by *J*, *K*, and CLK.

The time waveform of Figure 6-15 shows how the flip-flop responds to clock pulses. In this case, the flip-flop changes on the negative or trailing edge of the clock pulse. The time waveform shows response only when the PS and CLR are disabled. Note that a 1 at both *J* and *K* causes the output to toggle. This occurs at clock pulses 1, 2, 3, and 4. Triggering occurs only on the negative transition of the clock. The holding mode takes place

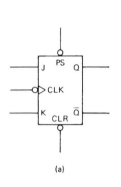

(a)

Operational mode	Inputs					Outputs	
	Asynchronous		Synchronous				
	PS	CLR	CLK	J	K	Q	Q̄
Asynchronous set	0	1				1	0
Asynchronous reset	1	0				0	1
Prohibit	0	0				1	1
Hold	1	1	0	0	0	No change	
Reset	1	1	0	0	1	0	1
Set	1	1	0	1	0	1	0
Toggle	1	1	0	1	1	1/0	0/1

(b)

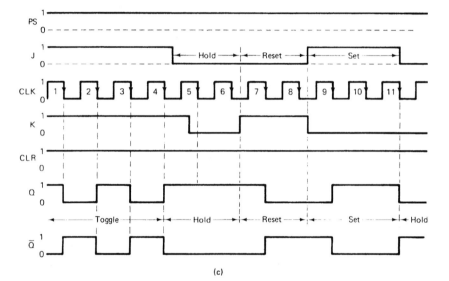

(c)

Figure 8-15. *JK* flip-flop: (a) symbol; (b) truth table; (c) waveforms

during pulses 5 and 6. *J* and *K* are both 0 at this time. Reset occurs during the time of pulses 7 and 8. The output goes low at the trailing edge of pulse 7. Setting occurs on the trailing edge of clock pulse 9. This continues for pulses 10 and 11. The waveform is in the holding mode after this point in time. Operation is continuous as long as power is applied and data supplied to the appropriate inputs. Several other variations of the *JK* flip-flop are now available on IC chips. Two or more *JK* flip-flops can be housed on a single chip.

SUMMARY

A flip-flop is a logic device with two or more inputs and two outputs. The outputs are the complements of each other. When the letter Q is used for the output it denotes the normal state. $\bar{Q}$ is the inverted output. The inputs are used to initiate a state change in the output. When an input is pulsed or triggered, it will send the output to a given state. The output will remain in this state even after the input returns to normal. This accounts for the memory characteristic of a flip-flop.

A bistable latch can be constructed with cross-connected NAND or NOR gates. Operation is dependent on the polarity of the voltage applied to its input. A 1 input applied to a NOR gate will produce a 0 output. With two of these gates cross-coupled, the output will turn on and latch when the input of one gate is 1 and the other is 0. Switching the polarity of the input causes the output to change its state. It remains in this state until instructed by the input to change. The output can be SET (S) or RESET (R) according to the polarity of the input. A NAND gate can be used to accomplish the same type of circuit, but the S and R outputs are reversed according to the input polarity. A latch is commonly used to debounce a switch.

An RS flip-flop can be constructed from two cross-coupled NAND or NOR gates similar to the bistable latch. The outputs are labeled Q and $\bar{Q}$, whereas the inputs are R and S. An active low RS flip-flop must have the S input 0 to produce a 1 at Q. An active high RS flip-flop must have a 1 at the S input to produce a 1 at $\bar{Q}$. A small circle attached to the input of an RS flip-flop symbol identifies an active low device and no circle denotes an active high device.

The clocked signal applied to a flip-flop is generally a rectangular-shaped series of waves. These clocked signals are defined as level triggering and edge triggering. In level triggering, the state of a clock changes value from 0 to 1 and carries out a transfer of data. Positive-edge triggering occurs at the leading edge of a pulse. Negative-edge triggering occurs at the trailing edge of a pulse. An edge-triggered flip-flop can have its data changed anytime, whereas level triggering can be changed only once at a specific level.

A clocked RS flip-flop has SET, RESET, and CLOCK inputs with Q and $\bar{Q}$ outputs. The inputs can be either positive- or negative-edge triggered. The outputs of an RS flip-flop are identified as Q_{n+1} and Q_n. Q_n is the status of the output before a clock pulse. Q_{n+1} is the output status after a clock pulse is applied.

A *D* flip-flop is a clocked *RS* flip-flop with an inverter connected across its inputs. The inverter causes the inputs to be opposites of each other. This causes the output to follow the value of the *D* input when it is triggered by a clock pulse. The *D* input is identified as data or delay. Negative- and positive-edge triggered *D* flip-flops are available.

The *JK* flip-flop is widely used in digital systems. This type of flip-flop has four possible input combinations. These are no change or hold, set, reset, and toggle. When the *J* and *K* inputs are both 1, this flip-flop does not produce an unpredictable output. The output changes states with each clock input. This is the toggle condition of operation. An SN7476 *JK* flip-flop has two asynchronous inputs called PRESET and CLEAR. These inputs override the *J*, *K*, and CLK inputs. When both asynchronous inputs are disabled, the synchronous inputs can again control the output. In the SN7476, the PS and CLR inputs are disabled by applying a 1 to them at the same time. This causes *J*, *K*, and the CLK to regain control of the output.

Chapter 9

Counter and Shift Registers

INTRODUCTION

Digital systems contain a large number of functional blocks that are used to perform specific operations. As a rule, these functional blocks do not perform operations until a specific command or clock pulse is given. This part of a system is considered to be synchronous, clocked, or a step-by-step procedure. Operations occur in sequence at a specific time. In order for this to occur, there must be data storage or some type of a memory capability. This all takes place in the functional block when it is made operational.

The primary element of a clocked system is a flip-flop. Flip-flops are capable of storing binary data and releasing it on command. Flip-flops are built into functional blocks and designed to respond to synchronous control signals. Counters and registers are examples of functional blocks that respond in this manner. The clock signal for this type of circuit is developed by an independent source and applied to the circuit components.

Counters and registers are fundamental elements of a digital electronic system. We take a closer look at these devices and their operation in this unit. In particular, we want to see how a variety of different counters respond and how registers are used to accomplish specific operations. These parts of a digital system are generally housed in packaged ICs.

DIGITAL COUNTERS

One of the most versatile and important logic devices of a digital system is the *counter*. This device, as a general rule, can be employed to count a wide variety of objects in a number of different digital system applications. While this device may be called upon to count an endless number of objects, it essentially counts only one thing, electronic pulses. These pulses may be produced electronically by a clock, electromechanically, photo-electrically, or by a number of other processes. The basic operation of the counter, however, is completely independent of the pulse generator.

Flip-flops

A flip-flop is the basic building block of a digital counter. The flip-flop was discussed in Chapter 8. Its operation is dependent on the status of the input and in some types on the condition of control elements. A *JK* flip-flop is widely used to form counter circuits. When this flip-flop is connected so that its *JK* inputs are at a 1, or high, level, it responds in the toggle mode of operation. This means that each clock pulse applied to the input causes the output to go through a state change. Toggling causes the flip-flop to have a divide-by-two function. The output complements itself with each input clock pulse. The first pulse causes the output to go high. The next pulse causes the output to go low. This action means that the output changes one complete time with two complete input pulses. A T-connected *JK* flip-flop has a divide-by-2 capability. This is fundamental in the operation of a binary counter.

The polarity of the trigger pulse is an important consideration in the operation of a flip-flop. Some flip-flops trigger on the leading or positive edge of the trigger pulse. Other flip-flops trigger on the trailing or negative going part of the pulse. Negative-edge triggering seems to be used more frequently in counters today.

Figure 9-1 shows the operation of a negative-edge triggered *JK* flip-flop. Note that the output has a short delay time before it is actuated. This is due to the propagation time of the flip-flop. This refers to the period of time needed to pass a signal through a flip-flop from the input to the output. As a rule, the delay is very small for a single flip-flop. Typical values are in the range of 20 to 30 ns. This permits 30-MHz signals to be processed. When several flip-flops are connected together in a counter, the propagation time increases significantly. Typical propagation delay times from input to output are in the range of 200 ns for a four-flip-flop counter. This permits clock signals of 5 MHz to be processed by a counter. Counters usually have longer propagation times than single flip-flops.

Binary Counters

A common application of the digital counter is used to count numerical information in binary form. This type of device simply employs a number of flip-flops connected so that the *Q* output of the first device

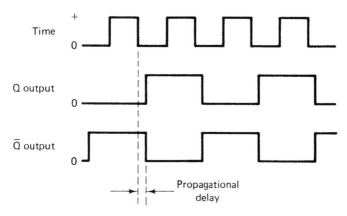

Time

Q output

Q̄ output

Propagational
delay

Figure 9-1. Timing diagram of a negative-edge-triggered *JK* flip-flop

drives the trigger or clock input of the next device. Each flip-flop therefore
has a divide-by two function.

Figure 9-2(a) shows *JK* flip-flops connected to achieve binary counting. This counter is commonly called a binary ripple counter. Each flip-
flop in this circuit has the *J* and *K* inputs held at a logic 1 level. Clock pulses applied to the input of FF-$_A$ cause a state change. This flip-flop circuit
triggers only on the negative-going part of the clock pulse. The output FF-
$_A$ therefore alternates between 1 and 0 with each pulse. A 1 output appears
at *Q* of FF-$_A$, for every two input pulses. This means that each flip-flop
divides the input signal by a factor of two. The *Q* output of each flip-flop
can then be considered as a power of two. The output of FF-$_A$ is 2^0, of FF-$_B$
is $2'$, of FF-$_C$, is 2^2, of FF-$_D$ is 2^3, and of FF-$_E$ is 2^4. Five flip-flops connected
in this manner will produce a count of 2^5, or 32_{10}. This is often called a
modulo-32 counter. The modulus of a counter is the number of different
states the counter must go through to complete its counting cycle. For a
modulo-32 counter the count is 00000 to 11111_2. The largest count that can
be achieved in this case is 11111_2. This represents 31_{10}. It occurs when all
flip-flops are 1 at the same time. The next pulse applied to the input clears
the counters so that 0 appears at all the *Q* outputs.

The counter of Figure 9-2, where the output of each FF serves as the
clock input for the next FF, is referred to as an asynchronous counter. This
is done because all FFs do not change states in synchronism with the clock
pulses. Only FF-$_A$ responds to the clock pulses. FF-$_B$ has to wait for FF-$_A$ to
change states before it is triggered. FF-$_C$, FF-$_D$, and FF-$_E$ must also wait for
a similar transition before they change. In effect, this means that there is a

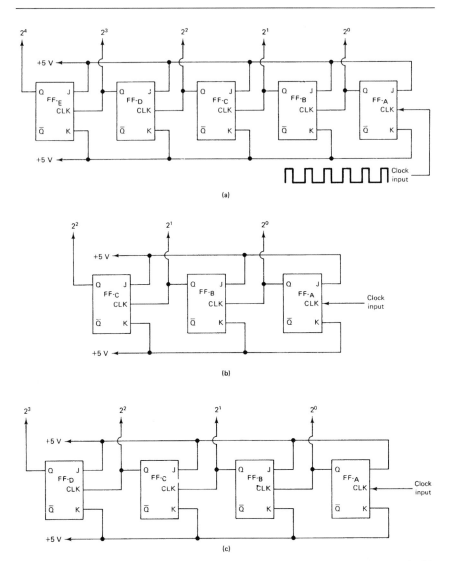

Figure 9-2. Counters: (a) binary counter; (b) octal counter (8 count); (c) hexidecimal counter (16 count).

delay between the responses of each FF. The delay is usually in the range of 20 to 30 ns. In some applications a counter with this amount of delay may be troublesome. Since a clock pulse applied to the input of the first FF causes a chain reaction, or ripple, to occur in the counter, this circuit is also called a *ripple counter*. The terms asynchronous counter and ripple counter are used interchangeably.

By grouping three flip-flops together as in Figure 9-2(b), it is possible to develop the units part of a binary-coded-octal (BCO) counter. Therefore, 111_2 is used to represent the seven counts, or seven units, of an octal counter. This is also called a modulo-8 counter. It counts from 000 to 111_2. The number of state changes that must take place is 8. Two groups of three flip-flops connected in this manner would produce a maximum count of $111\text{-}111_2$, which represents 77_8 or 63_{10}.

By placing four flip-flops together in a group as in Figure 9-2(c), it is possible to develop the units part of a binary-coded-hexadecimal (BCH) counter. This could also be called a modulo-16 counter. Thus 1111_2 would be used to represent F_{16} or 15_{10}. Two groups of four flip-flops can be used to produce a maximum count of $1111\text{-}1111_2$, which represents FF_{16} or 255_{10}. Each succeeding group of four flip-flops is used to raise the counting possibility to the next power of 16.

Binary counters that contain four interconnected flip-flops are commonly built on one IC chip. Figure 9-3 shows the logic connections of a 4-bit binary counter, or modulo-16 counter. When used as a 4-bit counter, the flip-flops will produce a maximum count of 1111_2 or 15_{10}. By disconnecting FF-$_A$ from FF-$_B$ and applying the clock to the input of FF-$_B$, we have a 3-bit, or modulo-8 counter. The Q outputs of flip-flops FF-$_A$ through FF-$_B$ are labeled 2^0, 2^1, 2^2, and 2^3, respectively.

Decade Counters

Since most of the mathematics that we use today is based on the decimal or base 10, system, it is important to be able to count by this method. Digital systems are, however, designed to process information in binary form because of the ease with which a two-state signal can be manipulated. The output of a binary counter must therefore be changed into a decimal form before it can be used by an individual not familiar with binary numbers. The first step in this process is to change binary signals into a binary coded decimal or BCD form.

Consider now the 4-bit binary counter of Figure 9-3(a). In this counter, 16 natural counts are achieved by the four flip-flops. To convert this counter into a decade, or modulo-10 counter, we must simply cause it to skip some of its natural counts. Notice the 16 natural counts listed below the binary counter. These read right to left.

A method of converting a binary counter into a decade counter is shown in Figure 9-3(b). In this method the first seven counts occur natu-

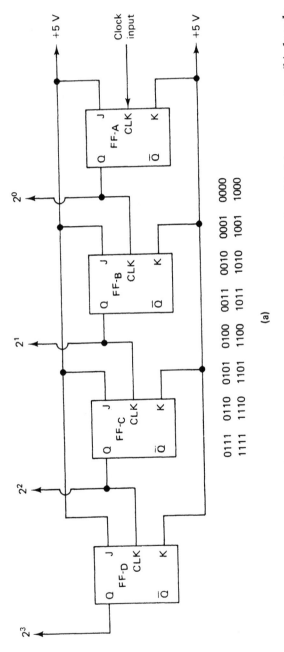

(a)

0111 0110 0101 0100 0011 0010 0001 0000
1111 1110 1101 1100 1011 1010 1001 1000

Figure 9-3. Binary counter to decade of BCD counter conversion: (a) four-bit binary counter; (b) decade, or BCD, counter

(Continued)

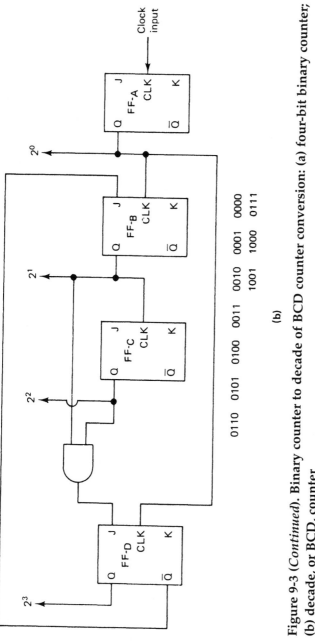

0110 0101 0100 0011 0010 0001 0000
              1001 1000 0111

(b)

Figure 9-3 (*Continued*). Binary counter to decade of BCD counter conversion: (a) four-bit binary counter; (b) decade, or BCD, counter

rally, as shown. Through these steps FF-$_D$ stays at 0. The Q output of FF-$_D$ therefore remains at 1 during these counts. This is applied to the J input of FF-$_B$, which permits it to trigger with each clock pulse. At the seventh count, 1 s appearing at the Q outputs of FF-$_B$ and FF-c are applied to the AND gate. This action produces a logic 1 and applies it to the J input of FF-$_D$. Arrival of the next clock pulse triggers FF-$_A$, FF-$_B$, and FF-c into the off state and turns on FF-$_D$. This represents the eight count.

When FF-$_D$ is in the on state, Q is 1 and $\bar{Q}$ is 0. This causes a 0 to be fed to the J input of FF-$_B$, which now prevents it from triggering until cleared. Arrival of the next clock pulse causes FF-$_A$ to be set to a 1. This registers a 1001$_2$, which is the ninth count. Arrival of the next count clears FF-$_A$ and FF-$_D$ instantly. Since FF-$_B$ and FF-c were previously cleared by the seventh count, all Os appear at the outputs. The counter has therefore cycled through the ninth count and returned to zero ready for the next input pulse. BCD counting of this type can be achieved in a number of ways. This method is quite common in IC devices today.

Figure 9-4 shows an IC BCD counter. The operation of this IC is essentially the same as the one just described. When this IC is used as a BCD counter, the output of FF-$_A$ must be connected to the BD input. Omitting this connection and applying the clock pulses to the BD input produces a five count. This permits the circuit to be changed to a modulo-5 counter. With this operation the counter is somewhat more versatile. The two NAND gates of this IC are used to set or reset the four flip-flops from an outside source. This counter triggers only on the negative-going part of the clock pulse.

Down Counters

Up until this point we have looked at counters that make an upward count. These counts progress from 0 upward and increase, or increment, the value. In some digital systems there is a need to count downward. A down counter counts from a higher value to a lower value. Before looking at a ripple down counter, let us examine the count-down sequence of a 3-bit down counter. Refer to Figure 9-5. Note that the count starts at its highest value and decrements to the next lowest value.

A JK flip-flop down counter is shown in Figure 9-6. This circuit is asynchronous and responds to the ripple effect, as did the up counters. The unique difference in its operation is determined by the polarity of its output. Note that the Q output of each flip-flop is connected to the clock input of the next flip-flop. The Q output of each flip-flop is attached to an

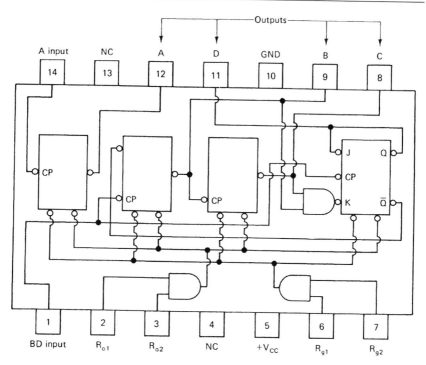

Figure 9-4. BCD counter IC 7490

Binary outputs

Figure 9-5. Count-count sequence

C	B	A	Base 10 count	
1	1	1	7	← The count starts here
1	1	0	6	
1	0	1	5	
1	0	0	4	
0	1	1	3	
0	1	0	2	
0	0	1	1	
0	0	0	0	
1	1	1	7	← Repeat count
1	1	0	6	
1	0	1	5	
1	0	0	4	
0	1	1	3	
0	1	0	2	
0	0	1	1	
0	0	0	0	

indicator that shows the response of the output. When FF-$_A$ toggles with the first clockpulse, it turns on. Output $\bar{Q}$ goes low at this time and turns on FF-$_B$. In the same manner, FF-c also turns on, and the counter shows 111_2, or 7_8. The second clock pulse causes FF-$_A$ to turn off. This action, however, causes Q of FF-$_A$ to go high. This does not alter the status of FF-$_B$. Since FF-$_B$ is unchanged, FF-c continues in the same state. The count has been reduced from 111_2, or 7_8, to 110_2, or 6_{10}. When the third pulse arrives, FF-$_A$ turns on and its $\bar{Q}$ output goes low. This action causes FF-$_B$ to change states and go low. The Q output of FF-$_B$ turns off, and the $\bar{Q}$ output of FF-$_B$, goes high. FF-$_C$, is unaffected by this action and remains in the high state. The count is now 101_2, or 5_{10}. Continuing in the same manner, the count is eventually reduced to 000 by the eighth input pulse. When the ninth input pulse arrives, it turns on all FFs for the counting sequence to be repeated.

Down counters are not as widely used as up counters. The primary application of a down counter is in situations where it must be known when a desired number of input pulses has occurred. In this application, the down counter is preset to a desired number. It is then permitted to count down until it reaches this number. When the number is reached, a logic gate detects the status of the counter and indicates the preset number of pulses that has occurred. This is often used to report the status of sequential operations in a complex digital system.

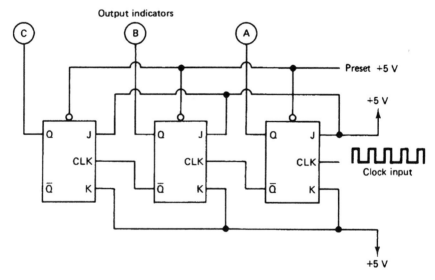

Figure 9-6. A 3-bit *JK* flip-flop down-counter

Synchronous Counters

Ripple counters require a certain amount of time for each flip-flop to change states in the operational sequence. This propagation delay causes the last FF to change states at a later time than the first flip-flop. This condition causes a serious limitation in the frequency of operation. In a synchronous or parallel counter, all flip-flops are triggered by the same input signal. This causes the flip-flops to change states at the same time. The outputs of each flip-flop will likewise change at the same time.

Synchronous counters are used to reduce the risks of false signal count production, to improve operational time, and to increase counting accuracy. As a rule, synchronous counters are faster, are more expensive, and use more power from the supply. These disadvantages are being reduced rather significantly with new IC designs. Synchronous counters are rather widely used in digital systems.

The operational requirements of a synchronous counter are somewhat different from those of a ripple counter. Each flip-flop of a parallel counter is driven by the same clock pulse. Logic gates are used to control the status of higher order flip-flops. Low-order (first and second) flip-flops must be on in order to produce triggering of the next flip-flop. The JK inputs of a flip-flop are controlled by signals that determine the status of a specific FF. In some circuits the preset (PS) and clear (CLR) inputs are used to alter the operation of the counter. A synchronous counter is usually more versatile than the ripple counter. A trade-off among cost, operational time, and power consumption must be decided when selecting a synchronous counter over an asynchronous device.

Figure 9-7 shows the circuitry of JK flip-flops connected in a 3-bit (modulo-8) synchronous counter. Each flip-flop has its input connected to the CLK input. FF-$_A$ has its JK inputs connected to the positive voltage source. A 1 applied to this input causes FF-$_A$ to toggle on the negative edge of each input pulse. FF-$_B$ has its JK inputs connected to the Q output of FF-$_A$. The status of FF-$_A$ determines if FF-$_B$ will toggle on a particular input pulse. If the Q output of FF-$_A$ is 0 before the clock pulse occurs, then J and K will be 0 and FF-$_B$ will not toggle when a clock pulse occurs. If the Q output of FF-$_A$ is 1 before the clock pulse occurs, then J and K will be 1, and FF-$_B$ will toggle when the clock pulse occurs. Similarly, FF-c will toggle only when the Q output of FF-$_A$ and FF-$_B$ are both 1 before the clock pulse occurs. These voltage levels are applied to a two-input AND gate. The output of the gate is connected to the JK input of FF-$_C$. It will change states only when FF-$_A$ and FF-$_B$ have the correct polarity. As a result, its output is controlled by the status of FF-$_A$ and FF-$_B$. The entire circuit will count in a binary sequence when the clock input is applied to each flip-flop. A parallel counter of this type requires more logic circuitry than a ripple counter.

The 74193 is a representative 4-bit synchronous binary up-down counter. Separate up-down clocks CP_u and CP_d are used to simplify the operation of this device. The outputs change states synchronously with a low-to-high transition of either clock input. The circuitry of this device is quite complex, as shown in Figure 9-8. The inputs are labeled D_0, D_1, D_2, and D_3, and the outputs are Q_0, Q_1, Q_2, and Q_3. The gate network attached to each flip-flop is the steering logic needed to control the counting direction. The counter also has an asynchronous reset and clear. A similar chip, the 74192, achieves synchronous BCD counting.

SHIFT REGISTERS

A register is a very important functional block in the operation of a digital system. Registers are primarily responsible for momentary storage of binary data. These blocks generally use flip-flops in their construction. Data supplied to the storage register are used to energize a particular flip-flop. When the flip-flop has been set to a 1 or 0, it will retain these data as *long as* energy is supplied to the system. The flip-flop can be cleared of stored data by actuating the clear or CLR input. The flip-flop will store data when it is triggered by a clock pulse. Storage registers are generally classified as synchronous devices.

A typical example of register operation occurs in the use of a digital calculator. Each time a number is entered on the keyboard, it appears in the display. The calculator must hold this number in a temporary memory register until told to do something else. If another number is entered, the first number shifts to the left one place. Two numbers are now held in the temporary memory register. The register must be capable of holding data in memory and be able to manipulate their positions. A shift register is generally used for this operation. It must hold data in storage and manipulate the data at the command of a clock signal. Shift registers can move data to the right or left of the decimal point. Data moved to the left represents multiplication by the base, or radix, of the system. For the decimal or base 10 calculator one place on the display represents a factor of 10. Moving data to the right represents division by the radix of the system.

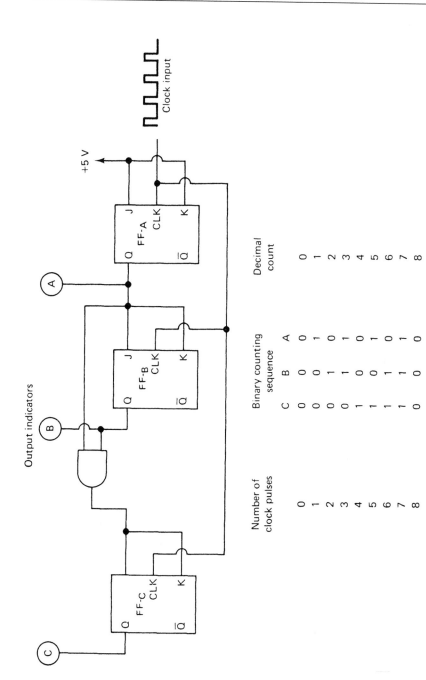

Figure 9-7. Three-bit (modulo 8) synchronous binary counter

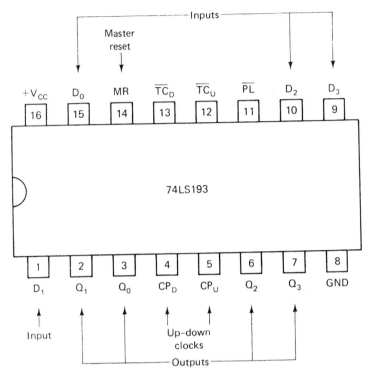

Figure 9-8. A 74LS193 4-bit synchronous binary up-down counter

Registers use the storage capability and shifting function of a flip-flop in their operations. The capacity of a shift register is based on the number of flip-flops in its construction.

A shift register is essentially a group of flip-flops that can be used to store binary data. There is generally one flip-flop for each-bit of binary data. The flip-flops are connected so that a binary number (data) can be entered and shifted into the register and in some cases be shifted out of the register. A group of flip-flops connected in this type of circuit must store data and shift it within the structure.

There are two ways of supplying binary data to a shift register. One method shifts data one bit at a time in a series configuration. This is usually described as a series-loading or series-input register. The second method loads data into the register all at the same time. This is called a parallel-loading or parallel-input register. Series- and parallel-loading shift registers are both used in digital systems.

Extracting or removing data from a shift register is another important characteristic of its operation. The simplest organization is a series-input/series-output (SISO) shift register. This register has data applied to one input and removed from one output. If the output is removed from each flip-flop of the shift register, it has a parallel output. There can be series-input/parallel-output (SIPO) or parallel-input/parallel-output (PIPO) shift registers. These circuit configurations permit data to be converted from series to parallel or from parallel to a series format.

Today shift register functions are often achieved by universal devices. They can be connected in a variety of different configurations. This permits the register to have many applications instead of one specific function. These registers are rather complex and are more difficult to construct with discrete flip-flops. New technology eliminates this restriction with multiple flip-flop stages being built on a single IC chip. We look at SISO, SIPO, PIPO, and universal shift registers in this presentation.

Series-loading Shift Registers

A series-loading shift register is shown in Figure 9-9. Four JK flip-flops are used in the construction of this register. It is designed to manipulate four bits of data at a time. Data are applied to the JK input of FF-$_A$ on the right side of the circuit. A NOT gate connected to the data input line feeds an inverted signal to the K input. When $J = 1$ and $K = 0$, the flip-flop will be set ($Q = 1$) with each clock pulse. With an inverter connected to the data input in this manner, the JK flip-flop responds as a D type flip-flop.

The output of a shift register can be developed from one flip-flop in a series configuration or from each flip-flop in a parallel configuration. The output usually has an indicator attached to it. An LED could be used for this function. When the LED is lit, it indicates a high, or 1, state; no light indicates low, or 0. The Q output of each flip-flop is also connected to the J input of the next flip-flop. The Q^* output connects to the K input of the next flip-flop. The clear or CLR inputs are commonly connected together. This permits the entire circuit to be cleared at the same time by an external signal.

Operation of an SISO shift register is controlled by two factors, the data input and the clock signal. Data applied to the input of the first flip-flop are transferred to the next flip-flop when each clock pulse occurs. To start this operation, it is desirable to clear the register of any previous data. The output of each internal flip-flop should be 0. This is accomplished by momentarily connecting the clear (CLR) input to 0. Then connect the CLR

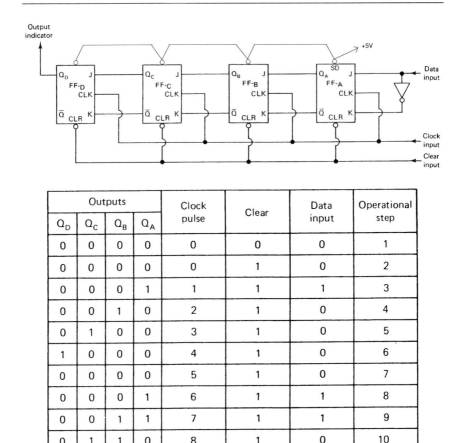

Outputs				Clock pulse	Clear	Data input	Operational step
Q_D	Q_C	Q_B	Q_A				
0	0	0	0	0	0	0	1
0	0	0	0	0	1	0	2
0	0	0	1	1	1	1	3
0	0	1	0	2	1	0	4
0	1	0	0	3	1	0	5
1	0	0	0	4	1	0	6
0	0	0	0	5	1	0	7
0	0	0	1	6	1	1	8
0	0	1	1	7	1	1	9
0	1	1	0	8	1	0	10
1	1	0	0	9	1	0	11
1	0	0	0	10	1	0	12
0	0	0	0	11	1	0	13

Figure 9-9. SISO shift register

input to a constant 1, or high-level source, for the remainder of the operation. Note this condition in the state table of Figure 9-9. Operational steps
1 and 2 on the right side of the table show the status of the register for this
condition.

An SISO shift register has the simplest construction of all shift registers. It has one input, one output, and several internally connected flip-
flops. The number of internal flip-flops determines the amount of data

that can be processed through the register. This type of shift register is primarily used to transfer data from one location to another in a digital system. The shift register of Figure 9-9 can manipulate four bits of binary data.

To demonstrate the operation of an SISO shift register, assume that we want to place a binary 1 into the input. This is represented by operational step 3 of Figure 9-9. The data input is 1, clear is 1 and clock pulse 1 occurs. This condition sets FF-$_A$ and causes the Q output of FF-$_A$, to be 1. The other flip-flops remain at 0. Step 4 shows the data input changed to 0 when clock pulse 2 occurs. This condition causes FF-$_A$ to be 0 and FF-$_B$ to be 1. The original 1 applied to the data input of step 3 is shifted one place to the left. Steps 5 and 6 show how this data is transferred through FF-c and FF-$_D$. The output of the shift register is connected to Q of FF-$_D$. The original 1 placed in the register will appear at the output after four clock pulses have occurred. These data have shifted through the four flip-flops during the operational cycle. Steps 8 through 13 show how the register will respond when additional data are applied to the input. In this case two consecutive is are placed into the register. These data shift through the internal flip-flops with each clock pulse change. They appear in the output after four clock pulses have occurred.

The SISO shift register of Figure 9-9 can be made into a SIPO register with a few simple modifications. An indicator, for example, must be attached to the Q output of each flip-flop in order for it to have a parallel output. This permits the output status of each flip-flop to be displayed. LEDs are attached to the Q output of each internal flip-flop. Figure 9-10 shows the circuitry of an SIPO shift register. The operation of this circuit is similar to that of the SISO register.

Assume now that we want to place a binary 1 into the SIPO shift register of Figure 9-10. To accomplish this operation, the register must first be cleared of all previous data. This is achieved by momentarily connecting the clear input to a low or 0 level. Step 1 of the status table shows this operation. The CLR must then be changed to a 1 (high level) before the shifting operation can begin. This is shown by step 2, where the data input is 1, the CLR is 1, and the clock input is 0. Nothing happens at this time because the clock input is 0. Step 3 shows the data input to be 1, CLR to be 1, and the clock pulse to be 1. This action causes the input data to be transferred to the output of FF-$_A$. Note that the Q output of FF-$_A$ is 1 and all other outputs remain at 0. Step 4 shows the same status for the data input, CLR, and clock. When clock pulse 2 arrives, it causes a 1 to appear at the Q output of FF-$_A$. The previous 1 in FF-$_A$ is shifted to the left and appears at the Q output of FF-$_B$. The register holds these data until the next clock pulse arrives. Step 5 shows a 0 applied to the data input. When clock pulse 3 arrives, the 0 appears at the Q output of FF-$_A$. The previous data are shifted to FF-$_B$ and FF-$_C$. Operational step 6 shows a 1 at the data input. When clock pulse 4 arrives, it shifts the previous data one position to the left and causes a 1 to appear at the Q output of FF-$_A$. The data from the most significant bit (MSB), or FF-$_D$ to the least significant bit (LSB), or FF-$_A$, is 1101. Operational step 7 causes the first data bit to be shifted out of the register. The process from this point on is repeated with different values of input data. Follow the operation of the register through the remaining steps. Note that data are transferred only when a clock pulse arrives at the CLK input. The data input of this register is loaded in a series operation.

The SIPO shift register of Figure 9-10 shifts data from the right to the left when a clock pulse arrives at its input. This operation is called a shift-left register. Shifting data to the left represents a multiplication operation. A position e in the place value of a number raises the power of the number by the base or radix of the numbering system being used. A binary number of 0001_2, or 1_{10}, would be 0010_2, or 2_{10}, after one position shift. In the first position the number value represents 2^0, or the number 1_{10}. When it shifts to the left by one position its place value changes to 2^1, or 0010_2. By this simple shifting operation we have raised the power of the original number by one. Each shifting step increases the power of the number by one. Shift-left registers are used to raise the powers of numbers. This can be used as a multiplication operation.

A SIPO shift-right register is shown in Figure 9-11. This register is similar in all respects to the shift-left register except the data input, clock, and clear inputs are on the left side of the register. FF-$_A$ is energized first. Data enter FF-$_A$ and shift to the right with each succeeding clock pulse. The display of data in a register is read with the MSB on the left and the LSB on the right. Data shifting to the right decreases the value of a number by the power of the base, or radix, of the numbering system. Essentially, this operation achieves division by the base of the numbering system. A number such as 1000_2, or 8_{10}, shifts to 0100_2, or 4_{10}, when it moves one place to the right. The original number, 8_{10}, when divided by the base of the numbering system (2), equals 4_{10}. Each shifting operation will cause this division operation to be repeated.

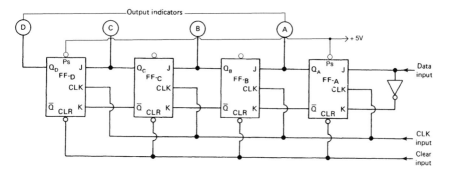

Outputs				Clock pulse	Clear	Data input	Operational steps
Q_D	Q_C	Q_B	Q_A				
0	0	0	0	0	0	0	1
0	0	0	0	0	1	1	2
0	0	0	1	1	1	1	3
0	0	1	1	1	1	1	4
0	1	1	0	1	1	0	5
1	1	0	1	1	1	1	6
1	0	1	0	1	1	0	7
0	1	0	1	1	1	1	8
0	0	1	0	1	1	0	9
0	1	0	0	1	1	0	10
1	0	0	0	1	1	0	11
0	0	0	0	1	1	0	12

Figure 9-11. SIPO shift-right register

Parallel-Input Shift Registers

In digital systems data comes in two different forms: series and parallel. Serial data is a stream of a single bits of digital information in the form of is and Os. Serial data can be applied to a shift register through one input. Parallel data are groups of related bits that all occur at the same time. A common size for parallel data is 8 bits. This group of data is called a *byte*. A byte of data represents a number greater than 0 and less than

256_{10}, or 2^8. Parallel data applied to a shift register necessitate an input for each bit of data. A byte requires 8 inputs. There data must all be supplied to the input of a shift register at the same time. Typical parallel input shift registers are designed to accommodate a byte of data.

A parallel loading input shift register must have an input for each bit of data it is designed to accommodate. The inputs of this type of shift register are uniquely different from those of a series-loading circuit. The preset (PS) input of a *JK* flip-flop, for example, is used to control the parallel input lines. The PS input is an asynchronous control. It, along with the clear input, overrides the synchronous inputs. As a rule, the asynchronous inputs are activated with a negative level signal. A circle attached to the PS and CLR inputs of a flip-flop symbol denotes negative-level control. Control signals for a parallel input must be inverted in order to achieve this operation.

A flip-flop diagram of a four-bit PIPO shift register is shown in Figure 9-12. This particular circuit shifts data from left to right. Notice that one flip-flop is used for each data bit. The Q and $\bar{Q}$ outputs of each flip-flop are connected to the *JK* input of the next stage. The output of each flip-flop is connected to an indicator. We will assume that the indicator is an LED. A clock input signal is commonly connected to each flip-flop. Data are applied to the PS input of each flip-flop. The polarity of this data must be at a negative level to initiate a state change in a flip-flop. Clearing the register of previous data is accomplished by a common negative level signal applied to each CLR input.

The status table of Figure 9-12 is used to show the operation of a PIPO shift register. When the circuit is first energized, it may have some random data in storage. This must be cleared before operation can begin. Clearing is achieved by momentarily connecting the common CLR line to a negative level, or 0. This operation resets each flip-flop and causes all the Q outputs to be 0. The CLR must then be switched to a 1 level to permit the operation to continue. Note operational steps 1 and 2. Step 1 shows some unwanted data stored in the output of the register. Step 2 shows the register cleared of unwanted data when the CLR input is switched to 0. Step 3 shows the CLR being changed to 1, data applied to the parallel inputs, and the outputs being set to the input data. Only negative-level data applied to the PS input will set the output of a respective flip-flop. Step 4 shows what occurs when the first clock pulse arrives. Note that the output data shift one position to the right. Steps 5, 6, and 7 show additional shifting when each clock pulse occurs. Step 7 indicates that the data have shifted

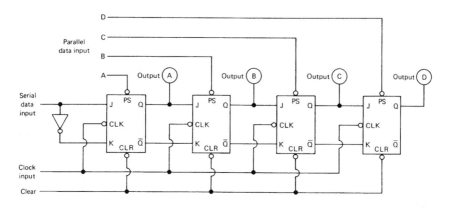

Operational step	Clear	Parallel data input				Clock pulse	Outputs			
		A	B	C	D		A	B	C	D
1	1	1	1	1	1	0	1	0	1	0
2	0	1	1	1	1	0	0	0	0	0
3	1	0	1	1	1	0	1	0	0	0
4	1	1	1	1	1	1	0	1	0	0
5	1	1	1	1	1	1	0	0	1	0
6	1	1	1	1	1	1	0	0	0	1
7	1	1	1	1	1	1	0	0	0	0
8	1	0	0	1	1	0	1	1	0	0
9	1	1	1	1	1	1	0	1	1	0
10	1	1	1	1	1	1	0	0	1	1
11	1	1	1	1	1	1	0	0	0	1
12	1	1	1	1	1	1	0	0	0	0

Figure 9-12. A 4-bit PIPO shift register

out of the register. With no input change the register would remain 0000, as indicated.

Assume now that additional data be applied to the inputs of our shift register. In this case, refer to step 8. Note that the input is 0011 and no clock is present. This sets the output so that it is 1100. Step 9 shows how the register will respond when a clock pulse arrives. The data in this case shift one place to the right. Steps 10, 11, and 12 show the data shifting to the right with each clock pulse. They shift out of the register in step 12. The register is 0000 at this point. Operation is complete until new data are applied to the parallel input.

Universal Shift Registers

Data manuals of digital ICs show that manufacturers produce a wide variety of shift registers. As a rule, this has changed to some extent with the development of a universal shift register. The 74194 4-bit bidirectional universal shift register is an example of new technology applied to this part of the IC field. The 74194 IC is a very adaptable shift register. It can achieve most of the shift register functions we have discussed in a single package. Data can be loaded with a series or parallel input. The 74194 IC can shift data to the right or to the left. The output can be connected in a parallel configuration or be derived from one flip-flop for a series output. Several 4-bit registers can be connected together to accommodate large amounts of data. This type of register can also re-circulate data from its last output stage back to the first input stage.

Figure 9-13 shows a partial listing of information from the Signetics Corporation TTL data manual. It shows a device description of the 74194, a block diagram, pin connections, data table, and some representative waveforms. This information is used to show how the shift register can be connected to achieve its different functions. The data sheet gives an overview of the operation of this device.

The logic diagram of the 74194 of Figure 9-13 is quite complex. It shows that the universal shift register is composed of four RS flip-flops. This means that the chip is a 4-bit device. It also shows some unusual gating circuitry controlling each flip-flop. This is needed to change the circuit to its different operational states. The parallel inputs, mode control, shift direction inputs, and clock signal are all controlled by the gating circuit. This permits the IC to be changed to different operational states. The 74194 is a rather unique shift register device.

Figure 9-14 shows the 74194 connected as a SIPO shift register. Notice that the mode control inputs are connected so that S_0 is 1 and S_1 is 0. This sets the device for a shift-right operation with each clock pulse. Shifting to the right is defined as a data change from flip-flop Q_A to Q_D. When data reach Q_D, they exit the register. The data are lost for future use.

Figure 9-14(b) shows the 74194 connected for SIPO shift-left operation. This is accomplished by changing the mode inputs so that S_0 is 0 and

This bidirectional shift register is designed to incorporate virtually all of the features a system designer may want in a shift register. The circuit contains 45 equivalent gates and features parallel inputs, parallel outputs, right-shift and left-shift serial inputs, operating-mode-control inputs, and a direct overriding clear line. The register has four distinct modes of operation, namely:

Parallel (broadside) load

Shift right (in the direction Q_A toward Q_D)

Shift left (in the direction Q_D toward Q_A)

Inhibit clock (do nothing)

Synchronous parallel loading is accomplished by applying the four bits of data and taking both mode control inputs, S_0 and S_1, high. The data are loaded into the associated flip-flops and appear at the outputs after the positive transition of the clock input. During loading, serial data flow is inhibited.

Shift right is accomplished synchronously with the rising edge of the clock pulse when S_0 is high and S_1 is low. Serial data for this mode is entered at the shift-right data input. When S_0 is low and S_1 is high, data shift left synchronously and new data are entered at the shift-left serial input.

Clocking of the flip-flop is inhibited when both mode control inputs are low. The mode controls of the SS4194/N74194 should be changed only while the clock input is high.

(a) Description

(b) Block diagram

Figure 9-13. Four-bit universal shift register 74194

(*Continued*)

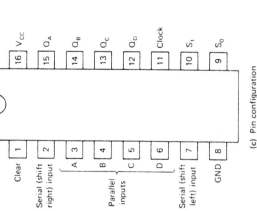

(Continued)

Inputs										Outputs			
Clear	Mode		Clock	Serial		Parallel							
	S_1	S_0		Left	Right	A	B	C	D	Q_A	Q_B	Q_C	Q_D
L	X	X	X	X	X	X	X	X	X	L	L	L	L
H	X	X	L	X	X	X	X	X	X	Q_{A0}	Q_{B0}	Q_{C0}	Q_{D0}
H	H	H	↑	X	X	a	b	c	d	a	b	c	d
H	L	H	↑	X	H	X	X	X	X	H	Q_{An}	Q_{Bn}	Q_{Cn}
H	L	H	↑	X	L	X	X	X	X	L	Q_{An}	Q_{Bn}	Q_{Cn}
H	H	L	↑	H	X	X	X	X	X	Q_{Bn}	Q_{Cn}	Q_{Dn}	H
H	H	L	↑	L	X	X	X	X	X	Q_{Bn}	Q_{Cn}	Q_{Dn}	L
H	L	L	X	X	X	X	X	X	X	Q_{A0}	Q_{B0}	Q_{C0}	Q_{D0}

H = high level (steady state)
L = low level (steady state)
X = irrelevant (any input, including transitions)
↑ = transition from low to high level
a, b, c, d = the level of steady state input at inputs A, B, C, or D, respectively
Q_{A0}, Q_{B0}, Q_{C0}, Q_{D0} = the level of Q_A, Q_B, Q_C, Q_D, respectively, before the indicated steady state input conditions were established
Q_{An}, Q_{Bn}, Q_{Cn}, Q_{Dn} = the level of Q_A, Q_B, Q_C, Q_D, respectively, before the most recent ↑ transition of the clock

(d) Truth table

Pin configuration:

Clear	1		16	V_{CC}
Serial (shift right) input	2		15	Q_A
A	3		14	Q_B
B	4		13	Q_C
C	5		12	Q_D
D	6		11	Clock
Serial (shift left) input	7		10	S_1
GND	8		9	S_0

Parallel inputs: A, B, C, D

(c) Pin configuration

Figure 9-13 (Continued). Four-bit universal shift register 74194

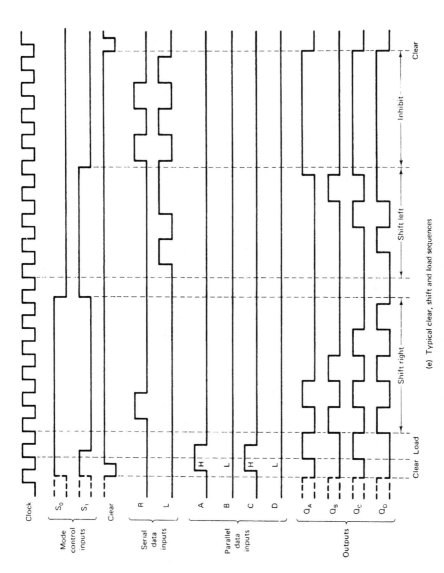

(e) Typical clear, shift and load sequences

Figure 9-13 (*Continued*). **Four-bit universal shift register 74194**

S_i is 0. The input data must be applied to the shift-left input. Shifting to the left is defined as a data change from flip-flop Q_D to Q_A. Data reaching Q_A exit the register and are no longer available for use.

Figure 9-15 shows the 74194 connected as a PIPO shift register. When a clock pulse occurs, the parallel data at inputs A, B, C, and D appear in the output display. The loading of input data occurs only when mode controls S_0 and S_1 are set to 1. The mode controls permit the circuit to have three operating conditions: shift-right, shift-left, and inhibit. The shift-right and shift-left serial inputs must be 0 to place data in the parallel inputs. No data are shifted when S_1 and S_0 are both 0. This is called the *inhibit mode* of operation. The mode controls are primarily responsible for the operation of the entire register. The CLR input clears the entire register to 0000 when it is 0. This control overrides all other inputs.

The universal shift register can be connected with another IC to accommodate larger amounts of data. Figure 8-16 shows two 74194 shift registers connected to form an eight-bit PIPO circuit. The parallel inputs are labeled A, B, C, and D for the first shift register and E, F, G, and H for the second register. All eight inputs are loaded into the register when a clock pulse arrives. This is achieved when S_0 and S_1, are both 1. The mode control can also be made to shift right or left according to the level of its input. Shifting right is achieved when S_0 is 1 and S_1 is 0. Shifting to the left is achieved when S_0 is 0 and S_1 is 1. The data-shifting operation is halted when the mode control is in its inhibit mode. This occurs when S_0 and S_1 are both 0.

The 8-bit shift register of Figure 9-16 is connected for re-circulating data. This is accomplished by connecting output Q_D of register 1 to the shift-right input of register 2. The Q_D, or H output, of register 2 returns to the shift-right input of register 1. Through this circuit configuration the data that exit register 2 return to the input of register 1. The data exiting output Q_D, or H, of register 2 are not lost through this type of circuit.

Universal shift registers are very useful digital devices. They respond to operations that require some form of temporary memory. They can also be used to delay information or data from one point to another. Shift registers are frequently used to achieve arithmetic operations. Microprocessors make extensive use of these devices. They are used to process information through various parts of the system. The universal shift register is a very important tool in the operation of a digital system.

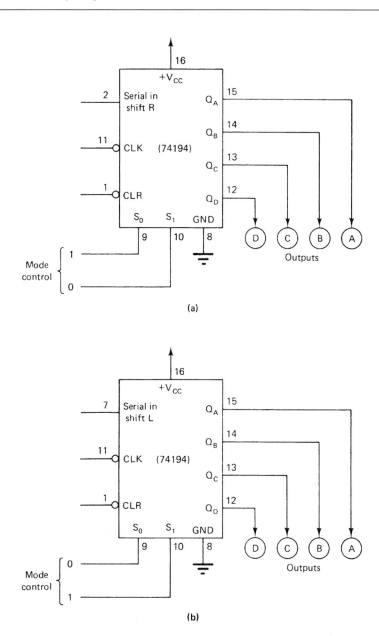

Figure 9-14. A 74194 connected as connected as a SIPO shift register: (a) shift-right position; (b) shift-left position

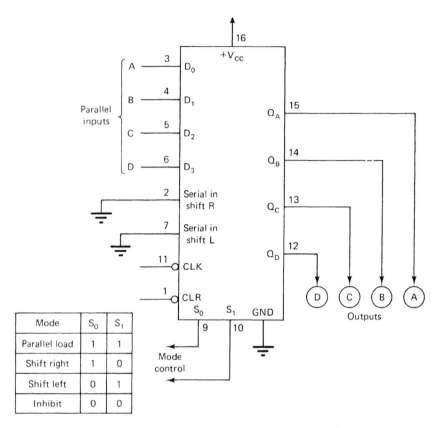

Figure 9-15. A 74194 connected as a PIPO shift right/left register

Mode	S_0	S_1
Parallel load	1	1
Shift right	1	0
Shift left	0	1
Inhibit	0	0

SUMMARY

Counters and shift registers are fundamental elements of a digital system. These elements perform operations when a specific command or clock pulse is given. They are considered to be synchronous or clocked or respond to a step-by-step procedure.

Digital counters are made up of flip-flops. *JK* flip-flops are widely used for this function. When this flip-flop is connected so that its *JK* inputs are 1, it responds in the toggle mode of operation. This means that each clock pulse applied to the input causes the output to go through a state change. Toggling causes the flip-flop to have a divide-by-two function.

Three flip-flops grouped together form a binary-coded octal coun-

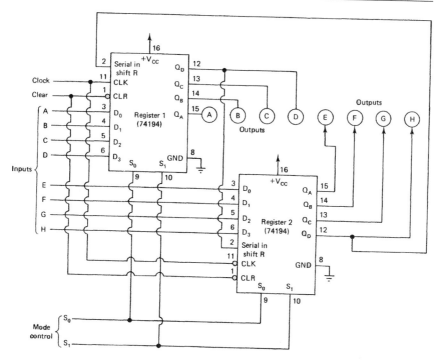

Figure 9-16. 74194 ICs connected as an 8-bit parallel-loading parallel-output shift-right register.

ter. This is called a modulo-8 counter. A modulo-8 counter goes from 000 to 111_2 and represents 7_8. When four flip-flops are connected together in a group, it is possible to develop the units part of a binary-coded hexadecimal counter. This is generally called a modulo-16 counter. Thus 1111_2 is used to represent F_{16}, or 15_{10}. Two groups of four flip-flops could be used to produce a maximum count of $1111\text{-}1111_2$, which represent FF_{16}, or 255_{10}.

Four flip-flops are commonly connected together to form a binary-coded decimal counter. In this counter six of the natural counts are skipped or omitted. An AND gate built into the counter produces a logic 1 from count 0111 and applies it to the J input of FF_{-D}. The next count turns off FF_{-A}, FF_{-B}, and FF_{-c} and turns on FF_{-D}. The counter shows a 1000. The next count turns on FF_{-A} to produce a 1001. The next count clears all the flip-flops, causing the counter to display 0000.

A binary down-counter counts from a higher value to a lower value.

This is accomplished by connecting the $\bar{Q}$ output of each flip-flop to the clock input of the next flip-flop.

Synchronous, or parallel, input counters trigger each flip-flop at the same time by the clock signal. This type of counter is considered to be synchronous because each flip-flop is triggered at the same time. Parallel input counters reduce the risks of false signal count production, improve operational time, and increase counting accuracy.

Shift registers are made of flip-flops. They hold data in memory and manipulate their positions. The data supplied to a shift register may be loaded one bit at a time in a series configuration or be entered all at the same time in a parallel-loaded circuit. Data may be removed one bit at a time from a series output or removed all at the same time from a parallel output. Shift registers are identified as SISO, SIPO, PIPO, and universal shift registers.

Chapter 10

Data Conversion

INTRODUCTION

The variation of electrical signals with respect to time can be either analog (continuous variation without discontinuities) or digital (having discrete steps with sudden discontinuities). Most physical quantities such as room temperature, air pressure, fluid flow, mechanical tension, and light intensity, vary continuously and are thus regarded as analog quantities.

Consider the fluid flowing through a mechanical flow valve shown in Figure 10-1. The flow values are of interest are 'no flow', 'normal', 'critical'. In this case either three separate binary-valued sensors for each condition would be required to monitor the three different flow conditions. Additional binary valued sensors would be needed if other flow values were of interest. Alternatively, use can be made of a suitable transducer which converts the fluid flow rate into an equivalent analog voltage or current. The word transducer is derived from 'trans-' meaning across and the word 'induce', meaning thereby that it is a device which can transfer some characteristic of a physical property by inducing equivalent change in another. This equivalent change could be in terms of some electrical quantity, such as voltage, current, or resistance. For example while monitoring the intensity of light use can be made of a light dependent resistor (LDR), one which changes the value of its resistance when the intensity of light falling on it changes.

Although most physical systems are largely analog in nature their monitoring and control is done using digital devices. Before most digital systems can act on analog inputs some type of transformation to the digital realm must occur. Similarly, in many cases the output of a digital system must be transformed back into the analog realm. This is shown in Figure 10-2.

Once we have the equivalent electrical quantity it needs to be converted into digital form for processing by a digital circuit. The device which handles this conversion is an Analog to Digital Converter or ADC.

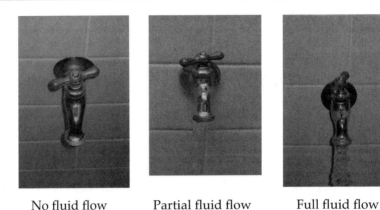

No fluid flow   Partial fluid flow   Full fluid flow

Figure 10-1. Variation of Fluid flow in a pipe

The ADC in the case of the flow system could for example generate three distinct voltage levels say 0V, 2.5V, 5V. This information would then be provided to the digital system for processing. Once processing by the digital circuit is complete, a digital output is produced. In cases where an analog output is needed, the digital signal must be converted into analog form. The device used for this is a DAC (Digital to Analog Converter). Finally, an output actuator is used for converting the equivalent analog voltage or current into the required analogous physical quantity.

Figure 10-3 shows an analog waveform. Notice the smooth variation of the analog voltages in the figure with no abrupt discontinuities in the magnitude at any time instant.

In Figure 10-3 an analog signal given is to be digitized, i.e., converted into a digital representation such as an equivalent binary number. The magnitude resolution or simply the resolution is the smallest change in magnitude which can be obtained following the analog to digital conversion. While the analog quantity is changing continuously, the electronic circuits which do the actual conversion to an equivalent binary number limit the speed of conversion. Thus, conversion from analog to digital is done only a regular time intervals, called the sampling period, with the value of the digital signal being maintained constant between the time intervals. This is shown in Figure 10-4 (a) using the dotted lines to indicate the value used for the entire time interval following the time instant the analog sample was taken. The final digitized waveform is shown in Figure 10-4 (b). Notice the abrupt changes in the values of the digital voltages.

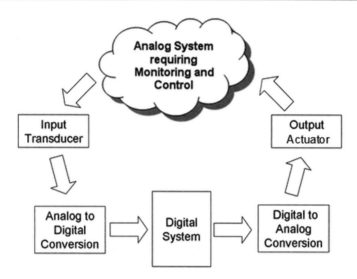

Figure 10-2. Converting analog information to digital and vice-versa for monitoring and controlling an analog system using a digital circuit

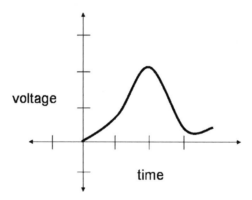

Figure 10-3. Analog signal

At some future time this data can be converted back from digital to analog form. This is shown in Figure 10-5(a). This may be done by using one location in each time interval of the digital waveform and connecting those points to create an equivalent analog waveform as is shown in Figure 10-5(b). The recreated analog waveform is shown using a solid line, whereas the original analog waveform using a dotted line. Notice that the recreated waveform differs from the original waveform. This is due to two factors – the number of samples taken per second, as well as

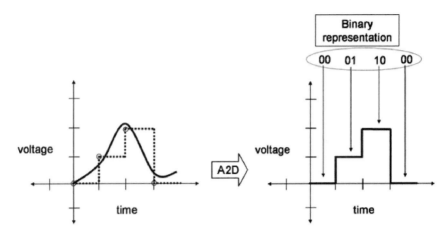

Figure 10-4. Creating a digital signal from an analog signal

the resolution of each sample encoded in binary format. When both the number of samples taken per second as well as the resolution is high, the digitized waveform can be used to accurately depict the original analog waveform.

The working of most Analog to Digital and Digital to Analog converter circuitry requires the use of high-speed and accurate electronic devices. This device will determine whether a given analog quantity is larger, smaller or equal to a given reference value. For the purpose of comparing analog quantities use is made of operational amplifiers, or op-amps. These are the analog equivalent of digital comparators used to compare multi-bit numbers.

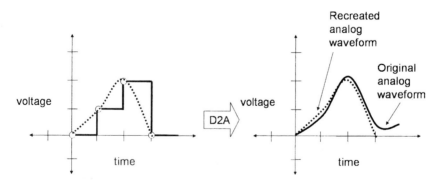

Figure 10-5. Converting an digital signal into an analog signal

OPERATIONAL AMPLIFIERS

Op-amps are multi-purpose two-input, one-output, analog devices. Op-amps have numerous operational features which make them well suited for use in analog-digital conversion circuitry including:

- stability with respect to temperature variations,

- high input impedance (so they do not alter the effective external electrical components connected to the input of the amplifier),

- low output impedance (so they do not alter the effective electrical load connected to the output of the amplifier),

- virtual grounding between the two input terminals if one of the input terminals is grounded, i.e. 0V (so that the voltage of the other input terminal would also be 0V)

- a very high gain (output/input), typically 10^5

- the ability configure the effective gain of the operational amplifier using external electrical components

The two input terminals of the operational amplifier are designated as (-) for the negative or inverting terminal; and (+) for the positive or non-inverting terminal. It is energized by a split power supply identified as $+V_{ref}$ and $-V_{ref}$. Typically the value of the source can be set to between ± 5-25V.

Operational amplifiers are used in two modes:

- open-loop—no feedback from the output of the amplifier is fed back into the input

- closed loop— feedback from the output of the amplifier is fed back into the input

Theoretically the open-loop gain of an op-amp is considered to be infinite. In effect whenever the input applied to the (+) positive input terminal is greater than that applied to the (-) negative terminal, the output of the operational amplifier, will rise to $+V_{ref}$ and if it is smaller the output voltage drops to $-V_{ref}$. The open loop configuration of an op-amp is shown in Figure 10-6.

Analog-Digital converters require that the digitized voltage output be similar in magnitude to the analog voltage at any given instant. This is clearly not the case when using the open loop op-amp configuration where the output is either $+V_{ref}$ or $-V_{ref}$ depending on the result of the comparison in voltages applied to the input terminals. So some modifications are needed to the open-loop op-amp circuit before it can be used with the analog-digital converters.

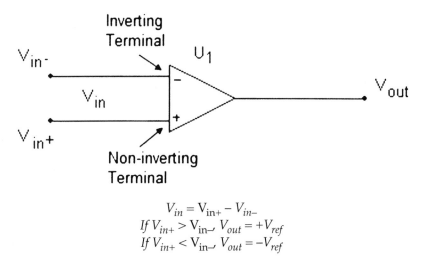

$$V_{in} = V_{in+} - V_{in-}$$
$$\text{If } V_{in+} > V_{in-,} \; V_{out} = +V_{ref}$$
$$\text{If } V_{in+} < V_{in-,} \; V_{out} = -V_{ref}$$

Figure 10-6. Open-loop behavior of an operational amplifier

Consider next the closed-loop configuration shown in Figure 10-7(a). It can be observed that, the output of the amplifier (V_{out}) is connected via a feedback resistor (R_f) to the inverting input terminal (-), and that the non-inverting input terminal (+) has been grounded. Owing to virtual grounding the voltage of the (−) terminal will become $0V$ as well.

The current in any branch of this circuit is determined by Ohm's law: $V = I \times R$ or $I = V/R$. Thus the current in the feedback branch, between points 'BN' of Figure 10-7(a) is V_{out}/R_f. Note also that in the circuit branch containing the input voltage (V_{in}) between points 'AN' of Figure 10-7(a) is V_{in}/R_{in}. As the input impedance of the op-amp is very large no current flows into the inverting terminal of the op-amp.

Note that the sum of currents at any given node (a point where multiple circuit branches meet), 'N' in Figure 10-7(a), equals 0. This is shown below:

$$\frac{V_{out}}{R_f} + \frac{V_{in}}{R_{in}} = 0$$

$$V_{out} = - \frac{R_f}{R_{in}} \times V_{in}$$

$$V_{out} = \frac{R_f}{R_{in}} \times (-V_{in})$$

$$\frac{V_{out}}{V_{in}} = - \frac{R_f}{R_{in}} = Closed\ Loop\ Voltage\ Gain$$

Thus, if Vin is positive the output Vout will be negative and vice versa.

As shown in Figure 10-7(a) the output of the inverting op-amp circuit is:

$$V_{out} = - \frac{R_f}{R_{in}} \times V_{in}$$

$$V_{out} = \frac{10,000}{20,000} \times (5)$$

$$V_{out} = - 2.5V$$

Since digital circuits normally process positive valued signals, the input voltage is often chosen to be negative, so that the output voltage can be a positive quantity.

Consider next the circuit shown in Figure 10-7(b) which is that of an op-amp in a non-inverting configuration, called a unity buffer or voltage follower. The voltage output is of this op-amp configuration is equal to the voltage input, but with a substantially higher current capacity. It is thus frequently used in electronic systems for providing higher current sourcing capabilities at the same voltage so it can be used to drive higher electrical loads. It is particularly useful for boosting the current sourcing

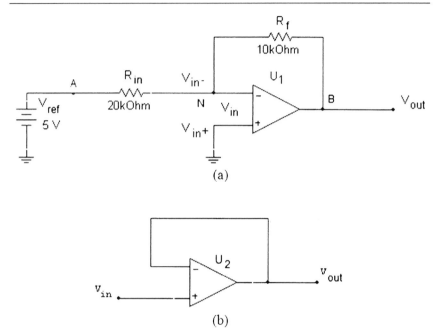

Figure 10-7. Closed-loop inverting configuration of operational amplifier

capabilities of external sensors and feedback elements connected to a digital system.

SAMPLE AND HOLD (S/H) CIRCUITS

When a continuously varying analog quantity is sampled (its magnitude recorded), the size of the memory where the magnitude is stored, and the speed of the conversion process determine the accuracy of the conversion process. It has been shown that in order to recreate a signal which has a maximum frequency component of 'f', it should be sampled at twice of this frequency, or at '2f'. For example, in order to successfully digitize the human voice whose frequency range is 20 Hz – 20 kHz, the signal should be sampled at 2 x 20 kHz, or at 40 kHz. Superior audio systems use samples at much higher frequencies, and also record the magnitude of the sample using multiple bits, typically between 8-24 bits.

Once the sampling frequency has been decided use is made of sample and hold circuits, similar to the one shown in Figure 10-8. In Figure 10-8 the op-amp U1 has unity gain, and responds as a voltage follower. This configuration tracks the variations in the input voltage, so that the applied analog input voltage appears on the output of U_1. It also has high current sourcing capabilities. Normally the switch J_1 is closed, so that this voltage is used to charge the capacitor C_1. At the instant of sampling, the switch J_1 is opened and voltage last recoded by the capacitor C_1 is used as input to the second voltage follower op-amp U_2. After the sample has been taken the switch J_1 is closed by the sample and hold control permitting the analog voltage to again charge up capacitor C_1 and prepare it for the next sample.

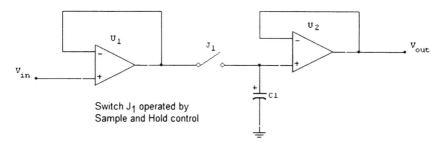

Figure 10-8. Sample and hold circuit

ANALOG TO DIGITAL CONVERTERS

(a) Flash or Parallel ADC

The flash ADC makes use of comparators for generating an equivalent digitized output. It does not require any clocks, and the instant an analog voltage is applied to the inputs, this voltage is compared using multiple comparators for generating an equivalent digitized output.

It is one of the fastest ADC available, but requires significantly more circuitry in the form of comparators. For a single bit ADC, $2^1 - 1 = 1$ comparators are needed. This is because for a 1 bit ADC all we have to do is perform one comparison to see if the analog input voltage is greater or smaller than a given threshold voltage.

For a 2-bit ADC, $2^2 - 1 = 3$ comparators are needed as shown in Figure 10-9 (a). In this case we need 4 different voltage levels corresponding to the 4 digital values possible shown in Figure 10-9 (b). Similarly, for a 3-bit $2^3 - 1 = 7$ are needed. As can be seen increasing the resolution of the ADC by 1 bit, requires the number of comparators to double. For a N-bit flash ADC converter, $2^N - 1$ comparators are needed in order to generate the appropriate digitized output.

Since the input analog voltage is being simultaneously applied to several comparators, it is possible that several of them simultaneously generate a logic high output. These comparator outputs are used as inputs to a priority encoder which determines the equivalent binary code corresponding to any given binary input.

The resistors used in conjunction with the comparators are selected so that the voltage increases uniformly in steps as the voltage is increased from the minimum to the maximum. As shown in Figure 10-9 the step size of the 2-bit flash is 1V, ranging from 0V to 3+V. Any voltage exceeding 3V such as 3.5V or even 4.5V would cause the comparators U_1, U_2 and U_3 all to generate a logic high output.

(b) Successive Approximation ADC

This is one of the most widely used types of ADCs as it has a very short conversion time from the application of an analog input voltage to the generation of equivalent digitized output. It consists of a memory register called the SAR (successive approximation register) having a fixed number of bits, a Digital to Analog converter and a comparator as shown in Figure 10-10(a).

At the start of an analog to digital conversion cycle, the highest order bit of the SAR is set, and the DAC generates an equivalent analog output voltage ($V_{out\,(DAC)}$) which is compared with the applied analog input (V_{in}). If $V_{in} < V_{out\,(DAC)}$, this means that the analog input is smaller than the equivalent voltage generated corresponding to the highest bit. In that case the SAR bit is reset, and the 2nd highest SAR bit is set, which is once again converted into an equivalent analog voltage by the DAC for comparing with V_{in}. This process continues until the lowest order bit in the SAR has been set and an equivalent voltage generated by the DAC for comparison with the analog input V_{in}.

The operation of a 2-bit successive approximation ADC is shown in Figure 10-10(b) where a digitized output corresponding to an analog input of 2.5 V is to be generated.

Each conversion cycle of the successive approximation ADC takes the exact same amount of time, since each bit of the SAR is set in turn, starting from the highest, and a voltage equivalent of it is compared with the analog input. Based on the result of the comparison SAR bits may remain set or be reset as the conversion cycle proceed, and is shown in Figure 10-10.

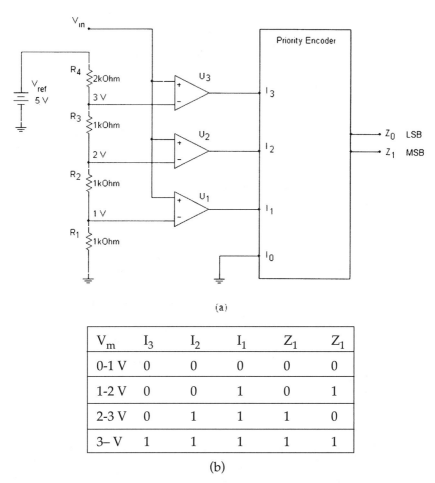

(a)

V_m	I_3	I_2	I_1	Z_1	Z_1
0-1 V	0	0	0	0	0
1-2 V	0	0	1	0	1
2-3 V	0	1	1	1	0
3– V	1	1	1	1	1

(b)

Figure 10-9. Two-bit flash ADC

ADC SPECIFICATIONS

(a) Resolution

The resolution or step size is the smallest magnitude which can be converted from digital to analog form by the ADC. For ADCs where the step size is large, the changes between the corresponding analog voltages between step sizes will be large as well, thus making the errors between the actual and theoretical values larger.

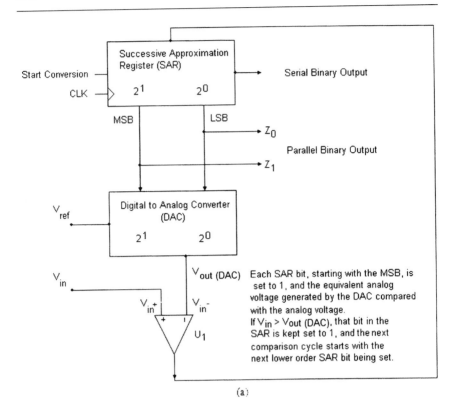

(a)

Step	SAR 2^1	SAR 2^0	DAV $V_{out(DAC)}$	Comparator U_1	SAR 2^1	SAR 2^0	Z_1	Z_0
1	1	0	2 V	Positive $V_{in+}>V_{in-}$	Set=1	0	1	0
2	1	1	3 V	Negative $V_{in+}<V_{in-}$	Set=1	Reset=0	1	0

$$V_{in} = 2.5 \text{ V}$$
DAC: $V_{ref} = 10V$ step size = 1 V
Final parallel binary output corresponding to $V_{in}=2.5$ V is $Z_1=1$, $Z_0=0$
(b)

Figure 10-10. Successive approximation ADC

(b) Quantization Error

A quantization error is the difference between the digitized value and the actual analog value. As the number of bits used for representing the magnitude increase, the quantization error reduces. There is always some difference between the two values, since any analog value can take on an infinite number of values between minimum and the full scale deflection value. It is specified as a +1 LSB (least significant bit) error, which indicates that the resultant digitized voltage could have a magnitude error of as much as the value corresponding to 1 LSB.

(c) Conversion Time

The conversion time is the time taken for the ADC to complete the digitization of any given analog signal. Smaller conversion times are desirable for high-speed ADCs. Certain ADCs as we have seen, such as the successive approximation ADC, have a fixed conversion time regardless of the analog quantity being converted. In contrast the speed of conversion in the case of a flash ADCs is determined only the underlying comparators, making it one of the fastest ADCs which are available. As the conversion time reduces, ADCs can take more samples and thus generate a much more accurate digitized version of a given analog waveform.

(d) Accuracy

The accuracy of the DAC depends on any underlying errors in the circuit components used in the DAC, and not on its resolution. The circuit components include power supplies, comparators, switches, resistors, etc, and is indicated as a percentage of the full scale (FS) reading, such as 0.02% FS. As the resolution of the ADC increases it is likely that the number of electronic components it uses will increase as well, and this may affect the accuracy of the overall system.

DIGITAL TO ANALOG CONVERTERS

Weighted Resistor DAC

The circuitry and working principle of a basic weighted resistor DAC is shown in Figure 10-11 (a)-(d). It makes use of op-amps configured to operate in an inverting configuration, along with suitable switches and resistors which have different weights selected so as to be inversely proportional to the weights of the particular bit position. Thus for a 2-bit

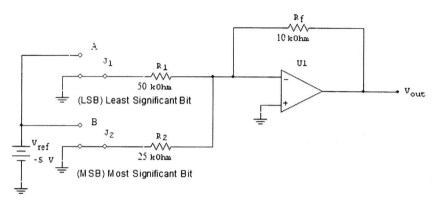

Input: B=0, A=0
$$V_{out} = 0V$$
(a)

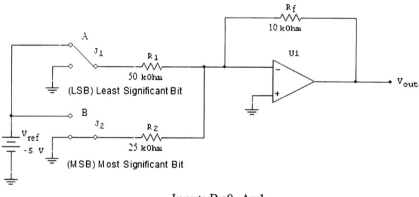

Input: B=0, A=1

$$V_{out} = -V_{ref} \times \left(\frac{R_f}{R_1}\right)$$

$$V_{out} = 5 \times \left(\frac{10,000}{50,000}\right) = 1V$$

(b)

Figure 10-11. Weighted Resistor DAC

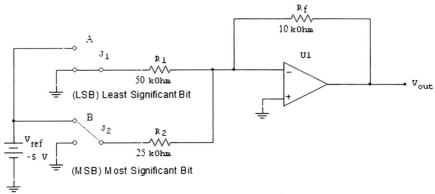

Input: B=1, A=0

$$V_{out} = -V_{ref} \times \left(\frac{R_f}{R_2} \right)$$

$$V_{out} = 5 \times \left(\frac{10,000}{25,000} \right) = 2V$$

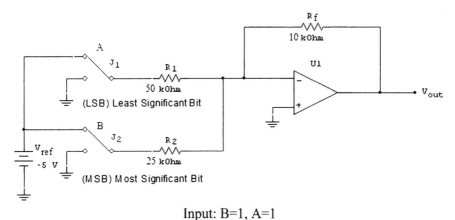

Input: B=1, A=1

$$V_{out} = -V_{ref} \times \left(\frac{R_f}{R_1} + \frac{R_f}{R_2} \right)$$

$$V_{out} = 5 \times \left(\frac{10,000}{50,000} + \frac{10,000}{25,000} \right) = 5x(0.2 + 0.4) = 5x(0.6) = 3V$$

(d)

Figure 10-11 (*Continued*). Weighted Resistor DAC

DAC, with the 1st bit having a weight of 1, and the 2nd bit a weight of 2; starting with the resistor for the 2nd bit having a resistance R of 25 kΩ, the value of the resistor for the 1st bit would be 2R or 50 KΩ. When this occurs the current which flows though the feedback resistor it generates a proportional amount of voltage that is double the value for each higher order bit as compared to the previous one. By selecting the resistors in this manner we can ensure that the values of the output voltage are weighted with binary multipliers corresponding to any given switch position, from the MSB (most significant bit) to the (LSB) least significant big.

The voltage V_{out} for the inverting op-amp circuit shown in Figure 10-11 is given below:

$$V_{out} = -V_{ref} \times \left(\frac{R_f}{R_1}\right) - V_{ref} \times \left(\frac{R_f}{R_2}\right)$$

$$V_{out} = -V_{ref} \times \left(\frac{R_f}{R_1} + \frac{R_f}{R_2}\right)$$

When the switches 'A' and 'B' are opened and closed the voltage output changes. Note that the weights of the resistors used in conjunction with the switches have been selected so that the values of the resistors double—with the resistor in the branch which serves as the MSB being the smallest. With the successive doubling of resistor values placed in the input branches from the MSB to the LSB, the current flowing in each branch is reduced by half.

Assuming the same voltage (V_{ref}) is applied in each input branch, the current flowing through the higher and lower order bit branch can be calculated using Ohm's law ($V=I \times R$) as:

- V_{ref}/R flowing through the higher order bit branch, and

- $V_{ref}/2R$ flowing in the lowest order bit branch.

This current flowing through the feedback resistor R_f, generates appropriately scaled outputs corresponding to different input switch arrangements. This is shown in Figure 10-11 (a)-(d).

One problem with this type of DAC is the significant difference in resistance values between that used in the MSB input branch, through that used in the LSB input branch, especially for multi-bit DACs. Manufacturing tolerances for resistance are in the ±10% range and for the LSB (which for say a 160kΩ resistor would be 16kΩ), this might exceed the resistor value itself used in the MSB (10kΩ of a 5-bit DAC, with resistances 10k, 20k, 40k, 80k, 160k in various input branches), giving rise to digital to analog conversion errors. Another problem is finding the precise value resistors which are needed for creating the weighted resistor circuit.

A staircase shaped waveform rather than a true analog waveform is generated whenever a DAC is used for converting a digital voltage into an analog one. As the number of bits representing the magnitude of the voltage is increased along with the conversion time, the staircase shaped waveform approximates an analog waveform.

R-2R Ladder DAC

As we have seen the weighted resistor DAC uses resistors which have different values, doubling in magnitude from the MSB to the LSB. It would simplify DAC circuit construction significantly if resistors of only a few different types were used regardless of the number of input bits. The R-2R ladder solves this problem making use of just two resistor values. When these resistors are connected in a ladder circuit, it results in a step-wise change in the value of the output voltage as each input bit is successively switched on.

Prior to looking at the R-2R circuit, a brief review of equivalent resistor calculations, and Thevenin equivalent circuit is provided, which will be helpful in understanding the operation of the R-2R ladder circuit.

RESISTANCE CALCULATIONS

When resistances R_1 and R_2 are connected in series, as shown in Figure 10-12 (a), the value of total series resistance, R_{series}, is equal to the sum of individual resistances.

$$R_{series} = R_1 + R_2$$

When series resistances are connected to a voltage source there is only one path for the current to flow through the individual resistances.

When resistances R_3, and R_4 are connected in parallel, as shown in Figure 10-12 (b), the value of the total parallel resistance, $R_{parallel}$ or $R_3 \mid \mid R_4$, is calculated using the formula below:

$$\frac{1}{R_{parallel}} = \frac{1}{R_3} + \frac{1}{R_4}$$

$$R_{parallel} = \frac{R_3 \times R_4}{R_3 + R_4}$$

It should be noted that the total resistance of a parallel circuit is always less than the smallest resistor in the circuit. When parallel resistances are connected to a voltage source there are multiple paths for the current to flow.

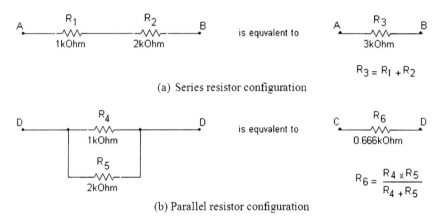

(a) Series resistor configuration

(b) Parallel resistor configuration

Figure 10-12. Series/Parallel resistor configurations

THEVENIN'S EQUIVALENT CIRCUIT

Any circuit, having multiple circuit elements, and with a load resistor R_L connected between any two terminals say, A and B, can be replaced by a single voltage source V_T in series with a single resistance R_T and the load resistance R_L. R_T is the resistance of the circuit measured between terminals A and B if the load R_L is removed and all voltage sources are replaced by their internal resistances (0Ω for ideal voltage sources). V_T is the voltage which is obtained across the terminals A and B if the load R_L is removed.

The process of obtaining the Thevenin equivalent circuit is shown in Figure 10-13 (a)-(f). The circuit whose Thevenin equivalent circuit is to be found with reference to terminals A and B is shown in Figure 10-13 (a). The resistance R_3 which is connected across terminals A and B is designated as the load resistor R_L in Figure 10-13 (b), and the rest of the circuit is disconnected from these two terminals. In Figure 10-13 (c) the equivalent Thevenin resistance R_T is found across the terminals A and B, after removing the load resistance R_L and also shorting out the voltage source V_1. This is shown in Figure 10-13 (d). For calculating the equivalent Thevenin voltage V_T across the terminals A and B, after removing the load resistance R_L, first the current (I) flowing in the series circuit is calculated in Figure 10-13 (e), and using this value of I the voltage drop across R_2 which is also the voltage across terminals A and B. Finally in Figure 10-13 (f) the voltage of Thevenin equivalent circuit is drawn with V_T, R_T and R_L connected in series.

With an understanding of series and parallel resistor combinations, and of obtaining a Thevenin equivalent circuit for a given electrical circuit, we proceed with the setup of the R-2R ladder. This is shown in Figure 10-14 (a)-(d), which correspond to the 4 different configurations based on the input switch arrangement of switches A and B. It can be seen that as the switches are opened and closed the output analog voltage changes as well.

DAC SPECIFICATIONS

(a) Resolution or Step size
The smallest change which can occur in the magnitude of the analog signal caused by a change in the magnitude of the digital signal is called the step size. For a DAC it is specified as the magnitude of the LSB.

(b) Full-scale output
The maximum analog output voltage produced by the DAC, corresponding to the highest magnitude of the input digital signal is called the full-scale output. With a larger number of steps and if each step has a higher magnitude, the total full-scale output can be increased.

$$Percentage\ resolution = \frac{step\ size}{full\ scale\ DAC\ output} \times 100$$

(c) Percentage Resolution
The magnitude of the step size as compared to the full scale output

of the DAC expressed as a percentage is often used to determine the error which can exist in the DAC. This is termed as the percentage resolution.

(d) Accuracy

The accuracy of the DAC is determined by the accuracy of the electronic components such as voltage sources, comparators, resistors, etc., which are used to built its circuitry. It is expressed in terms of the full scale DAC output and often as a percentage.

(e) Full-scale error

The maximum deviation of the output of a DAC from its expected value, as expressed as a percentage of the full scale value is called the full-scale error.

(f) Linearity error

The maximum deviation in the step size of a DAC as compared to the expected step size is called the linearity error.

(g) Monotonicity

A DAC is said to be monotonic if its analog output increases as the digital input is increased in steps through its complete range of operation from minimum to full scale.

(h) Settling time

The settling time is the time taken by a DAC to settle within $\pm 1/2$ step size of the full scale value starting from the minimum value at the start of the digital to analog conversion process. This reflects the maximum time which can be taken in the DAC process, when the digital input is changed from all 0s to all 1s.

(i) Offset error

When the digital input is set to minimum the analog voltage should ideally be at a minimum as well. If this is not the case owing to underlying inaccuracies in the components used to build the DAC, an offset error exists, which will in effect be added to the magnitude of the step size, thereby increasing the voltage (if the offset error is positive).

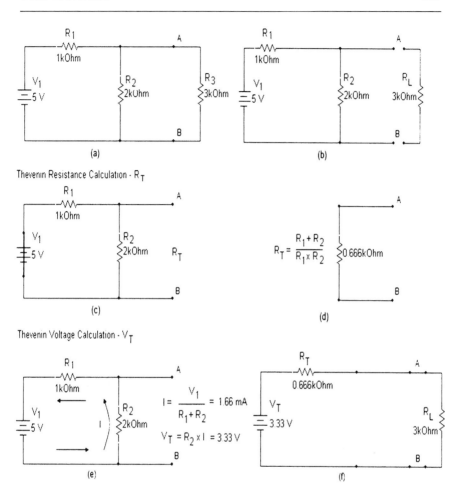

Figure 10-13. Thevenin equivalent resistor and voltage source calculations

SUMMARY

Quantities which can change their magnitude continuously are termed as analog quantities, whereas the ones which exhibit discrete jumps in their magnitude are termed as digital quantities. Transducers are used to convert these physical quantities into electrical signals. Sample and hold circuits are used to take a snapshot of an analog quantity so that it can be converted into an equivalent binary number.

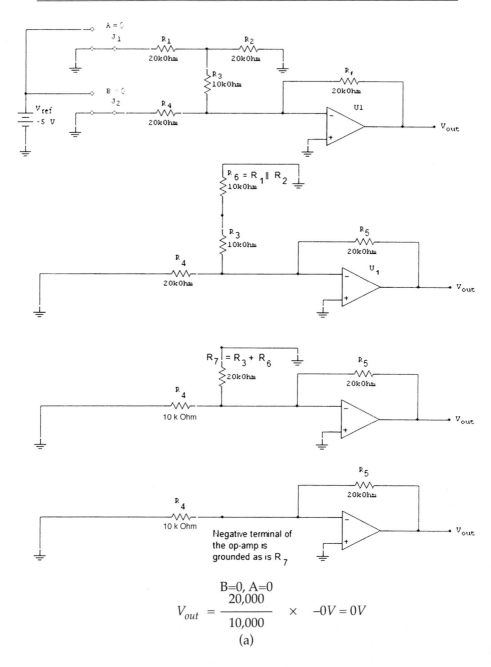

$$V_{out} = \dfrac{\overset{B=0, A=0}{20,000}}{10,000} \times -0V = 0V$$

(a)

Figure 10-14. R-2R Ladder DAC

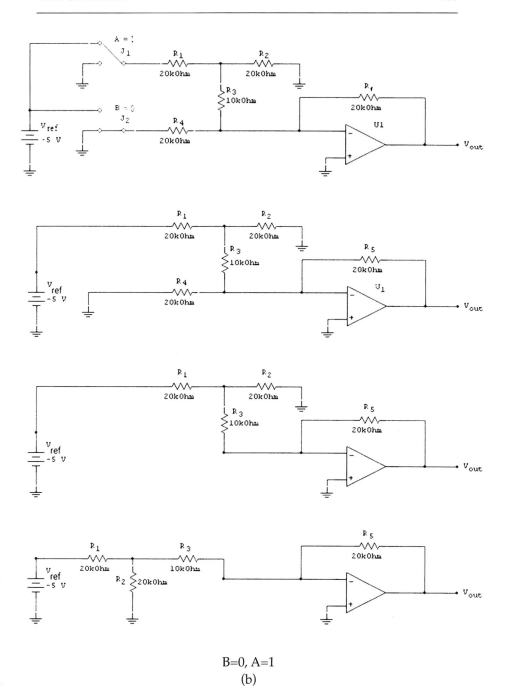

B=0, A=1
(b)
Figure 10-14. (*Continued*) R-2R Ladder DAC

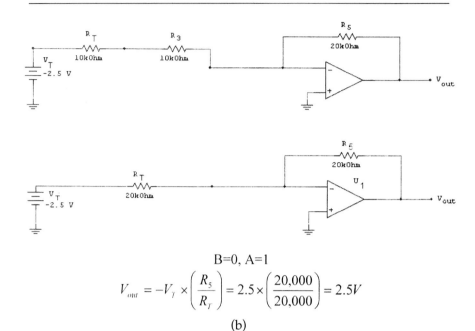

$$V_{out} = -V_T \times \left(\frac{R_5}{R_T}\right) = 2.5 \times \left(\frac{20,000}{20,000}\right) = 2.5V$$

(b)

Figure 10-14. (*Continued*) R-2R Ladder DAC

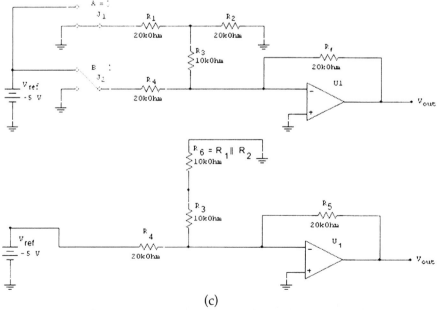

(c)

Figure 10-14. (*Continued*) R-2R Ladder DAC

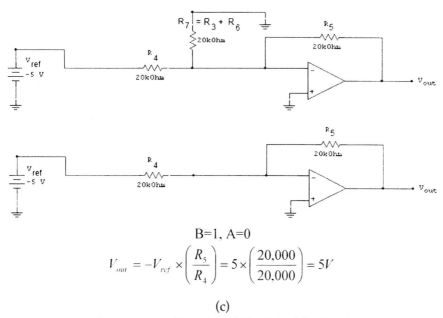

$$V_{out} = -V_{ref} \times \left(\frac{R_5}{R_4} \right) = 5 \times \left(\frac{20,000}{20,000} \right) = 5V$$

(c)

Figure 10-14. (*Continued*) R-2R Ladder DAC

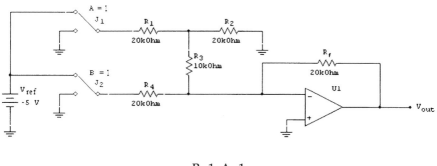

B=1, A=1

$$V_{out} = 7.5V$$

(d)

Figure 10-14. (*Continued*) R-2R ladder DAC

An ADC assigns a binary number to a given analog quantity. As the number of bits used to represent an analog quantity increases its resolution increases, and smaller changes in the analog quantity can be represented. Flash or parallel ADC use operational amplifier comparators and a priority encoder for generating an equivalent binary number to an applied analog input signal. The successive approximation ADC has a constant conversion time and finds extensive use as a general-purpose ADC.

A DAC converts a binary number into an equivalent analog voltage. A N-bit DAC takes a potentially infinite range of analog values between the minimum and the full scale, and sets up 2^N discrete voltage points. These serve as an approximation of the analog voltage corresponding to a given binary number. Use is made of weighted resistors for generating equivalent amounts of current corresponding to position of the bit in a given number. The highest-order bit contributes the most current and also generates the most output voltage. Each successive bit contributes half the current and the voltage to the output. The R-2R ladder arrangement is preferred while setting up in DAC circuits since it uses resistors which have exactly two values, regardless of the number of bits being converted into equivalent analog signals.

Chapter 11

Advanced Digital Concepts

Most of the digital devices in use today contain complex circuitry, containing thousands of combinational and sequential logic elements. In fact the arrangement of these elements, termed as architecture, is done so as to maximize the space, power, material, speed, efficiency requirements. One of the major advances in digital circuitry was accomplished using 'tri-state devices' and 'bus architecture' making it possible to share data.

Consider the digital circuit given in Figure 11-1. For each input device that connects to the digital circuit a number of input lines are connected to the circuit. Here it has been assumed that the device, say a microphone, mouse click, or a keyboard key press signal is being encoded into digital format, and is then transmitted in binary format to the digital circuit. With each new device, the digital circuit needs additional input connections. Once the number of input connections are exhausted no additional input devices can be connected, without first disconnecting some other. This is also the case for the outputs which the digital device can generate. It can be seen that the bottleneck is the number of input and output connections which are available for connecting to input and output devices respectively.

BUSES

In order to get around the problem of limited input/output connections, the 'bus architecture' was developed. This is shown in Figure 11-2.

Notice that the inputs, outputs, and the digital circuit itself, share a common bus. A bus is simply a group of wires used for transferring data in the form of logic signals from one digital device to another. As can be seen in Figure 11-2 the number of input and output lines is the same per input/output device. Voltage values assigned to unused lines are simply not used. Since any input device may decide to transmit data to the digital circuit at any given time using the bus, this can lead to unpredictable voltage on the

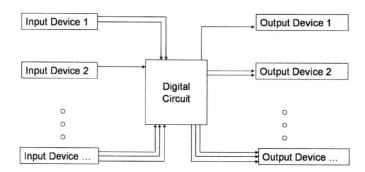

Figure 11-1. Input/output connections of a digital circuit

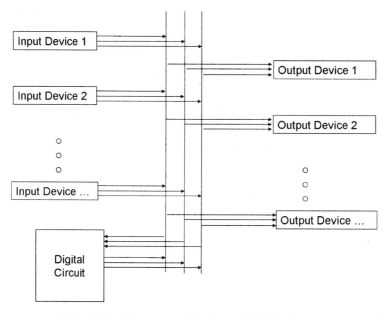

Figure 11-2. Digital circuit communicating with I/O devices using a data bus

line. Such a situation is termed as a 'bus fight' and is to be avoided.

As can be seen synchronization between the digital circuit and the input/output devices is needed. This is provided by 'enable' signals which the digital circuit generates in turn and polls each input device. Only one device is enabled and data is then transmitted from the input device to the digital circuit. Similarly, the digital circuit may enable an output device and then transmit data to it using the bus. This is shown in Figure 11-3.

While this setup solves the synchronization problem, one issue still remains—that of blocking voltage signals on the bus from reaching devices other than the ones it has been sent to. With various logic levels existing on the bus at any given time, all the devices connected to the bus can sense this voltage, whether the data is intended for the device or not. Electrically isolating the devices which are not the intended recipients of the signals on the bus would solve this problem. This is where tri-state devices play a key role in the proper functioning of the bus and of the devices which are connected to it.

TRI-STATE LOGIC DEVICES

In addition to the two state of logic low and high state, tri-state devices can be placed in a 'high-impedance' state. When in a high-impedance state the output of the device is electrically isolated from the input. Thus any device which needs to be disconnected from another device can be connected using tri-state devices. The symbol for a tri- state device is shown in Figure

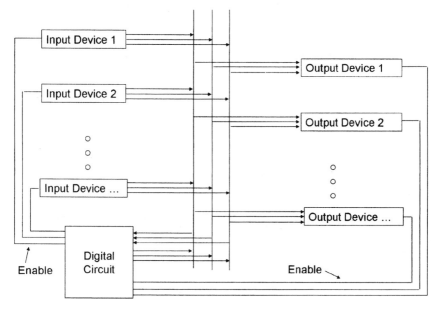

Figure 11-3. Coordinating data transfer using enable signals on the data bus

11-4(a) and the truth table is given in Figure 11-4(b).

Within the bus architecture discussed so far it is possible for input devices to send requests for data transfer (which may be regarded as an 'interrupt') to the digital circuit. This is shown for input device 1 in Figure 11-5. The digital circuit will send an acknowledgement back to the input device, and enable the device, as well as the associated tri-state buffers for transmitting the data to it.

In case several devices are requesting transmission of information to the digital circuit, some sort of priority is used to determine which interrupt has higher priority. In a computer system for example, a relatively slow device such as a mouse has a much lower priority than that used by the real time clock used.

CENTRAL PROCESSING UNITS (CPUS) AND MICROPROCESSORS

CPUs perform arithmetic and logic operations based on instructions ('programs') which are read in and interpreted ('decoded') into a binary number, which can then be processed. These include instructions to add numbers, to jump to different program locations, to store or retrieve values, and so on. CPUs include some very high speed storage registers for temporarily holding data while it is being processed, as well as primary

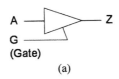

(a)

G (Gate)	A (Input)	Z (Output)	
0	0	High Impedance	Output, Z, disconnected from input, A
0	1	High Impedance	
1	0	0	Output, Z, connected to input, A
1	1	1	

(b)

Figure 11-4. Conventional circuit symbol and Truth Table for a tri-state device

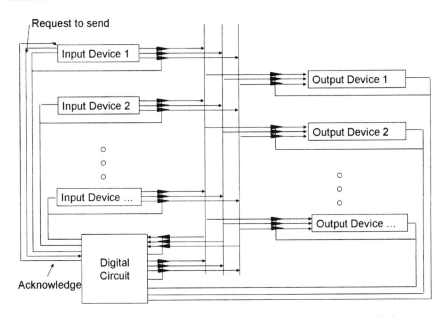

Figure 11-5. Using tri-state devices on the data bus for isolation

storage ('cache'). Input/output ('I/O') capabilities are also part of CPU.

The CPU controls the sequence in which operations occur, such as reading in of data from input devices or of programs from other storage media, including hard disks, instruction decoding, writing of data to storage, and output. These functions were performed on several different chips. Owing to advances in computer technology it is now possible to integrate all these different functions on a single 'chip' or block of material with miniaturized circuit elements, and is termed as the microprocessor.

Programs which are meant to run on a microprocessor consist of a series of instructions (add, subtract, branch, etc.) for performing an operation and certain operands (the number on which the operation is possibly being performed). Each instruction which a particular microprocessor and associated circuitry can process is termed as the Instruction Set Architecture (ISA) and is called an op code ('operational code'). Each op code is a binary number and the instruction set specific to each microprocessor. So for example the op-code for adding two numbers using a Pentium chip would be different from that used for a AMD Athlon chip. As can be imagined programming in using binary numbers can be quite hard, and error prone owing to the likelihood of incorrectly entered machine codes. Not only must the proper logic of the program be developed, each instruction must then be decoded into binary. Some instructions are multi-step. For example, addition of two numbers may first require each to be fetched and stored in the registers on the microprocessor, and then perform the actual addition. Troubleshooting machine code is excruciatingly hard. In addition to the use of locating errors in the program logic the binary number equivalents of each instruction have to be isolated.

The main components of a microprocessor are shown in Figure 11-6. Note the timing and control unit determines the sequence in which operations must occur in the processor. Use is made of the Interrupt controller and the I/O controller for connecting to external devices. Programs are fetched from memory making use of the bus controller which sets the proper memory address on the address bus, and fetches the data using the data bus. The program counter keeps track of the instruction being executed using the instruction register and decoder generates the proper sequence of machine code cycles corresponding to any instruction.

If for example two numbers to be added these may be placed in the temporary register and the result stored in the Accumulator after performing the proper operation in the ALU. If during this operation the number is negative, zero, or too large appropriate flags or bits are set in the flag register.

An alternate to machine code is to write the program using assembly language, which are mnemonics corresponding to each op code. These are somewhat easier to understand than, for example the add operation may simply be represented by the mnemonic ADD. One thing which should be noted about assembly language is that while it is much easier to understand by humans, it cannot be processed by the microprocessor, which is expecting the operation and operands to be binary numbers (in 16-, 32-, 64- or even 128-bit format based on the microprocessor). So the next step after writing a program in assembly language is to 'assemble' it, whereby all the mnemonic codes are decoded into machine language.

MEMORY SYSTEMS

Nearly all digital systems call for the rapid storage and retrieval of digital information. The amount of information to be stored ranges from a few million bits in a pocket calculator to several hundred trillion bits in a computer system. The technology of digital memory is perhaps the most

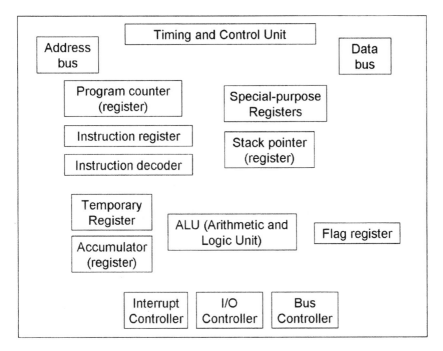

Figure 11-6. Main components of a microprocessor system

rapidly changing sector of the digital electronics field. In the last few years speed, storage capacity, and reliability have increased in overwhelming proportions. Improvements in memory-device capabilities will undoubtedly play a very important role in the future of digital electronic systems.

Electronics memory refers to a method of storing and retrieving digital information. Storage, or writing, is the process of placing digital information in memory. Retrieving, or reading, is the process of recovering memorized digital information. Memory involves both processes.

The memory section of a digital system has been a problem for some time. Other functions of the system have improved so rapidly that less time is now required to store and retrieve digital information. The memory section must be capable of accepting information and processing it through the system very rapidly with perfect accuracy. This involves such things as access time, word length, and storage capacity.

Each type of digital memory device is generally quite unique in its construction and internal operation. The basic operating principles of these devices are usually the same for the memory section of a digital system. An understanding of basic operating principles is essential in the study of individual memory devices.

Each memory device requires several different types of input and output lines to perform its operation. The response of the device is determined by control of its input/output. It must respond to a number of control operations in order to function:

- Select the address in memory that is being accessed for a read or write operation.

- Select the read or write operation to be performed.

- Supply input data to be stored in memory during the write operation.

- Retain the output data coming from memory during the read operation.

- Enable or disable the memory so that it will or will not respond to the read-write command.

Before a microprocessor can run a program, it must be loaded from secondary storage (hard-disk, CD, floppy, tape drive) to primary storage (memory). In secondary storage read or write operations take a considerably long time, typically several milliseconds. The read/write mechanism in secondary storage must be mechanically positioned prior to data being read in or written to the device. In contrast such read or write operations are significantly faster typically in the microsecond range or better when performed in the memory.

The primary storage consists of:

- Read-only Memory or ROM and permits only reading from it, not writing to it. Most recent versions of ROM permit a limited number of write operations as well.

- Random-access Memory or RAM, so named as it takes approximately the same amount of time to retrieve or write data regardless of where it is located.

- Cache, which is a small, very high-speed memory area located within or close to the microprocessor.

ROM

In ROMs the data storage is permanent, so the data is preserved when power is removed from the device. In essence it is an electrical equivalent of a look-up table or matrix, as in a spreadsheet. Data is arranged in cells, and identified uniquely by the corresponding row and column in the table. Fusible links are used to connect a voltage line to the output. If the fusible link is blown, then the voltage is sensed as a ground voltage or a 0, else it is sensed as the supply voltage or a 1. Thus a binary value can be stored at any location in the table. This is shown in Figure 11-7.

When the memory chip is enabled and the address of a particular row provided, the decoding circuit makes the corresponding line at the supply voltage high. This supply voltage travels across the selected row and at each intersection of a row and column examines whether the fusible link has been blown or not. If the fuse is not blown the supply voltage can travel down along the column to the input side of the tri-state buffers. When the ROM chip is enabled the tri-state buffers permit the voltage appearing on their input to be connected to the data output. Hence the supply voltage would appear on the output as well. If the fusible link is blown then 0V will appear at the input of the buffer corresponding to that column, and will be transferred to the output. As shown in the diagram, if

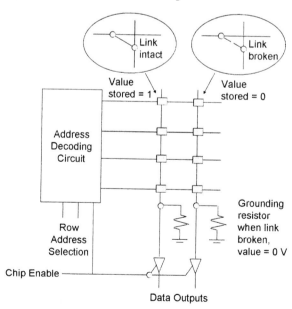

Figure 11-7. Address decoding circuit for a ROM

the top row is selected the data read out will be 10_2.

ROMs are useful in computer systems which on start up, require initialization instructions. These are stored in ROM BIOS (Basic Input Output System), and this helps the computer identify and test the major components before transferring control to the particular operating system in use, such as Windows XP, loaded on the computer. This process is also called the boot-up or bootstrap process. It would not be possible unless initial instructions were stored in a location that would be initiated at startup. The microprocessor on boot-up loads these instructions starting at a predefined memory location in ROM, using the RAM for storing data temporarily.

There are several types of ROMs. This includes:

- Programmable ROMs or PROMs which can be programmed only once.

- Erasable PROMs or EPROMs which permit erasing of stored program in the device using ultra-violet rays.

- Electrically erasable EPROMs or EEPROMs which permit erasing of the program stored in the device by using special voltage signals. This requires additional circuitry for performing the erase operation, which increases the size of the device, thereby reducing the amount of data stored per unit area.

- Flash drives are now available, which combine the features of EPROMs and EEPROMs, making high storage density and electrical erasure of programs possible. A representative flash drive is shown in Figure 11-8.

RAM

RAM is form of volatile memory, which means that these devices lose any stored data if power to the IC is removed. They thus need to be constantly powered to ensure data storage. The primary use of RAM is to temporarily store microprocessor based program instructions and the data needed by the programs. For example, information regarding the cursor or mouse on the screen needs to be updated constantly while a computer is on, as also information about any programs which are running.

There are two primary kinds of RAM. This includes:

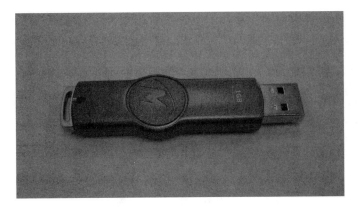

Figure 11-8. Flash (Thumb) drive

- Static RAM or SRAM, where the memory cells are composed of flip-flops for storing information.

- Dynamic RAM or DRAM where the memory cells are composed of capacitor based circuitry. In DRAMs the capacitors holding the value need to be refreshed with the stored data value at regular intervals (approximately 2 ms) or the value stored gets drained (lost).

DRAM circuitry is complex owing to the refresh cycles, it consumes more power, and is slower; whereas SRAM is the memory of choice, despite the fact that it requires more space per data bit stored. Its power consumption is very low, making it suitable for use in portable devices such as cell phones, digital cameras, calculators, etc. The memory card for a digital camera is shown in Figure 11-9. SRAMs and DRAMS use both row selector decode circuitry, as well as column selector decode circuitry in order to identify any particular cell location where a bit value has to be stored.

With regards to computer systems, when the term memory is used it generally refers to a RAM. In practice memory is most likely to be achieved with SRAM. These are soldered onto circuit boards which can then be plugged into a computer. When the RAM chips appear on only one side of the circuit board, the packaging is said to be Single Inline Memory Module or SIMM. When the RAM chips are placed on both sides of the circuit board, the packaging is said to be Dual Inline Memory Module or DIMM. As can be imagined DIMMs hold significantly more memory than SIMMs and most modern day computer systems use this packaging. SRAMs are packaged using a printed circuit boards for use as computer memory. Figure 11-9 shows a secure digital (SD) card for use with a digital camera.

Newer technologies have made it possible to read/write data not just on every clock pulse, but on the rising and falling edges of the clock pulse. This doubles the data which can be accessed. These are called Double Data Rate or DDR SRAM DIMMs. Typical DIMMs sizes range from 64MB to 2GB.

Hard Disk Drive (HDD)

A hard disk drive is designed with a very large magnetic storage capacity in mind, reaching into several hundreds of GB and into the TB (tera byte) range. It is a non-volatile storage device, consisting of data stored on several rotating disks called platters contained within a sealed unit. The disks rotate at several thousand (RPM) revolutions per minute, typically 7000 to 10, 000 RPM.

Figure 11-9. RAM used in a digital camera

The smallest unit of data storage on a hard disk is a sector, typically 512 bytes. Two or more sectors make up a cluster, which is the amount of space a file needs for storage, typically 4kB to 64kB. These sectors and clusters are arranged in concentric circles called tracks on the top and bottom of each platter. The vertical collection of a set of tracks from all the platters on a hard disk is called a cylinder. Data is read from and written to multiple tracks simultaneously. Random access times for the data stored on a HDD, ranges from 5 ms to 15 ms.

The platters made of a non-magnetic material, are covered with a fine layer of iron oxide, which is easily magnetized when exposed to a current carrying conductor. This is due to the phenomenon that current flowing though a conductor generates a magnetic field which has a certain magnetic orientation (South or North). The specific magnetic orientation is used for denoting the presence of zeroes and ones, and can be changed by reversing the direction of current flowing through the conductor.

A read/write head is located at the end of an actuator arm within the HDD, and is positioned less than 10 micrometers to several nanometers, on each side of all the platters. Individual read/write heads are used to retrieve data from and to write data to the tracks located at the top and bottom of each platter. Data is read by detecting the magnetic orientation of the recorded data, by the direction of current induced in the read head. A representative hard disk drive is shown in Figure 11-10.

HDD control electronics, integrated into the disk drive, include the disk controller which allows the microprocessor used with a digital system such as a computer or camera, to schedule disk read and write operations, specify type of data encoding schemes, compression or storage procedures. It may also include SMART (Self monitoring, Analysis and Reporting Technology) standard for informing users about possible disk failures. A host adapter is used to connect the host, such as a digital computer, to storage devices such as a HDD. ATA (Advanced Technology Attachment), also called IDE (Integrated Drive Electronics) is the standard parallel interface used for connecting disk drives to computers. Newer disk drives support the (SATA) Serial ATA bus standard and offer very high speed data transfers over serial links, reaching up to 3 Gbit/s and beyond.

Figure 11-10. Hard Disk Drive

GENERAL PURPOSE COMPUTER SYSTEMS

Earlier computer designs were hardwired systems which could not perform any other function than for which they had been created. They were in a sense pre-programmed, to perform a specific operation or sequence, such as adding numbers, or calculating the trajectory of a missile. Such fixed purpose devices were expensive to maintain. With the advent of the 'stored program' computer it was possible to run a variety of programs on the same fixed hardware.

Each program written in a high-level language, which would be compiled, that is, converted into machine code which the microprocessor could process and execute. High level languages include Java, C++, FORTRAN Formula Translator or FORTRAN, and Common Business Oriented Language or COBOL, High level computer languages are closer to human talk, but has a fixed number of keywords, constructs, constants, which form part of the set of programming language. While the op codes for different microprocessors may differ considerably, the high-level programming language remains essentially the same. This is because compilers are written for specific hardware architectures and translate the program into binary which can be processed on the microprocessor.

The microprocessor controls all the operations of a digital computer. A simplified block diagram of a microprocessor based computer system is shown in Figure 11-11.

The flow of data in a computer system occurs in the forward direction. Starting with the input controller, the data moves through the microprocessor, and out through the output controller. Programs are fetched from storage and each line of a program is executed in turn. Programs may make use of input or stored data for generating output, which may be directed to the output devices or to storage. The operation of the entire system is synchronized by a clock, which creates a very stable, periodic,

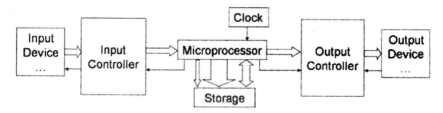

Figure 11-11. Block diagram of an earlier computer system

square shaped waveform. Control signals from the microprocessor manage various controllers such as that for the input, storage and output.

As can be observed, all the data entering the system goes though the microprocessor and this in many cases creates a bottleneck in the system, especially when a very slow device is transmitting data. In general the microprocessor can perform only a single operation at a given time, so while data is being received from a keyboard for example, a much faster device such as a hard-drive will have to wait for the microprocessor to complete processing the keyboard data.

Direct Memory Access or DMA controllers permit direct transfer of data between input devices and designated storage space, which can be processed by the microprocessor as needed. This is shown in Figure 11-12. The use of DMA greatly speeds up operations in the computer.

Figure 11-13 shows a motherboard, which is the main circuit board used inside a modern computer system. While the microprocessor remains the 'brains' of the operation it is the support chipset which determines the actual rate at which data communications can occur. The Northbridge chipset performs all the high speed data transmissions such as those which occur between the microprocessor and the memory. On the other hand, the Southbridge chipset performs data communications between relatively slow peripheral devices, such as the keyboard and the system.

MICROCONTROLLERS

Just as a microprocessor combines all the functions of a CPU on a single chip, a microcontroller goes a step further and combines all the func-

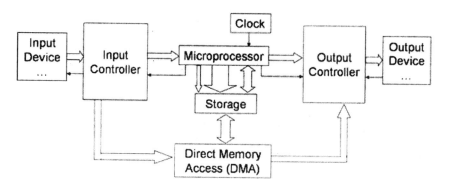

Figure 11-12. Block diagram of an advanced computer system

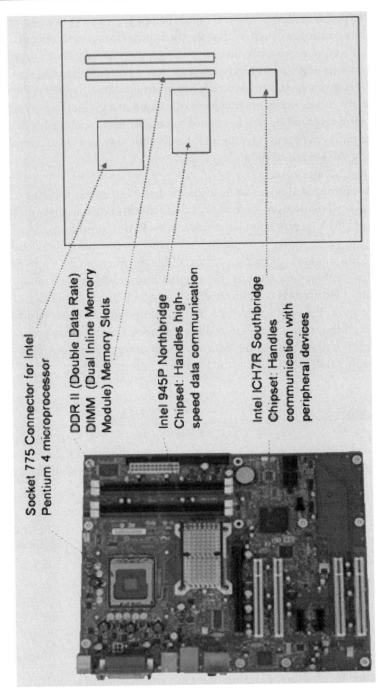

Figure 11-13. Main board of modern computer system

tions of an entire computer on a single chip. It thus has as a CPU, internal clock, storage (both ROM and RAM), Input/output control capabilities, and could even include power regulation components needed for circuit operation. This is shown in Figure 11-14.

Microcontrollers are a remarkable advance in the field of digital electronics. These devices make it possible to control a wide variety of systems, ranging from car ignition, temperature control, washing machines, ATMs, security systems, digital signal processing audio/video applications, microwave ovens, and numerous others. Figure 11-15 shows a representative microcontroller

The control code or program for the microcontroller may be written in assembly language, in which case it must be assembled and then downloaded to the storage on the microcontroller. Other alternatives include using an interpreted language, such as BASIC (Beginners All purpose Symbolic Instruction Code), wherein the program is written in a higher-level language, for processing on a per line basis by the microcontroller. It is then downloaded to the microcontroller. Each line of the program is interpreted, i.e., converted into machine language for the microcontroller to process and execute. This process of interpreting each line of the program slows down the overall execution time, and as such interpreted programs run slower than compiled ones. It should be noted that prior to download the program it should be checked for proper syntax, keywords or constructs. Most microcontrollers will not permit the

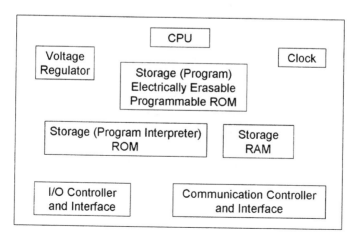

Figure 11-14. Main components of microcontroller

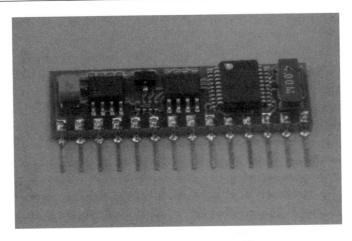

Figure 11-15. Microcontroller

code to be downloaded if it contains such errors. However, as yet there are no checks for logic errors in the program. Such errors are commonly identified by running the program and applying sample inputs (called test vectors) to activate all portions of the program code.

PROGRAMMABLE DEVICES

PLDs refer to the class of digital devices which have user-assigned logic functions. PLDs are available which contain both combinational and sequential logic features, making them suitable for a wide range of applications having low power, size and high-speed requirements. Early versions of PLDs included Programmable Array Logic (PAL) devices which contained a PROM array, implementing a sum-of-products, and output logic. The PALASM (PAL Assembler) design software was used for converting Boolean expressions into binary fuse patterns. These functioned similar to a lookup table based on the input applied to the device. Generic Array Logic (GAL) is a reprogrammable extension of PAL devices, thereby making them suitable for prototyping.

Complex PLD (CPLD) devices contain several interlinked PAL devices and are equivalent to several thousand or more logic gates, all on one integrated circuit. CPLDs are configured to perform specific logic functions by programming these devices using an industry standard JEDEC (Joint Electron Device Engineering Council) file format. These files are generated by computer programs called logic compilers. The source code for the logic compiler is written using a HDL (Hardware Description Language) such as PALASM or ABEL (Advanced Boolean Expression Language).

SUMMARY

In digital systems buses refer to multiple tracks of electrical connections which are used to transfer multiple bits of information between devices. Tri-state devices, which may have a logic Low, logic High, and High Impedance output state, are used in conjunction with buses for isolating devices which are not are not being used for data transfer.

Memory refers to a method of storing and retrieving digital information. Writing is the process of placing data in memory. Reading is the process of recovering data from memory. A memory location refers to an address where data is stored. This address must be selected to write or read data. Static memory is achieved with solid-state devices such as bipolar transistors and MOS transistors. Static random-access memory, or RAM, retains memory as long as power is maintained. Dynamic RAMs are volatile memory devices.

Read-only memory, or ROM, is designed to hold data that are either permanent or rarely altered. The process of placing data in memory is called programming. Some ROMs have data burned into memory by destroying fusible links to selected memory cells. Masked ROMs have selected memory cells programmed with photographic masks during the manufacturing operation. ROMs are nonvolatile memory devices.

Programmable read-only memories or PROMs, are user-programmed memory devices. PROMs can be altered only one time. They are nonvolatile memory devices. Erasable programmed read-only memory, or EPROM, can be programmed by the user. It can be erased and reprogrammed as often as desired. A program entered in EPROM memory is nonvolatile. It can be erased by exposing the chip to ultraviolet light. Electrically erasable PROMs are altered electrically. They represent nonvolatile memory when a program has been entered in memory.

General purpose computer systems can be programmed with high-level programming languages, with data in the system flowing in the forward direction. It uses input/output controllers, microprocessor, storage devices and clock for processing signals from input devices and generating outputs.

Microprocessors feature high-speed data processing capabilities. They run programs which can be used to process data inputs and generate outputs.

Microcontrollers are in essence a computer on a chip. They have input-output ports, arithmetic and logic processing capabilities, on-chip volatile and non-volatile memory, clock, buses for transferring data.

Appendix A

Electrical and Electronic Safety

Safety can never be stressed enough while working with electrical and electronic equipment. Since the voltages in use with electronic equipment are typically 5V or less, similar to what one may experience while handling a flashlight, it is easy to disregard standard safety practices.

Ensuring the safety of personnel and equipment is very important in the laboratory or the workplace. Many dangers are not easy to see. For this reason, safety should be based on understanding basic electrical principles. Common sense is also important. The physical arrangement of equipment in the electrical lab or work area should be done in a safe manner. Well designed electrical equipment should always be used. Electrical equipment is often improvised for the sake of economy. It is important that all equipment be made as safe as possible. This is especially true for equipment and circuits that are designed and built in the lab or shop.

The following points should be noted with regard to the use of safety.

FOR PERSONNEL

- Electrical Shock: Humans are in general good conductors of electricity. Electricity can cause burns, and when it flows through heart cavity it can cause the natural electrical rhythm to be disrupted.

- Be aware of what is happening around you, and if possible do not work alone.

- Know the location of the main breaker or fuse box in the room.

- Physically disconnect the power cable while working on high power equipment.

- Do not wear metal objects such as bangles, necklaces or chains while working with electrical systems, as contact though them could cause a severe electrical shock.

FOR MEASURING EQUIPMENT

- Examine the power cabling of any equipment used for measuring high voltages or currents.

- Use grounded power cables for equipment, especially those which require 120V AC or higher.

- Ensure that all power is switched off while making connections to the circuit. The best way to ensure this is by physically removing the power cable to the circuit.

- Discharge any capacitors prior to use in a circuit. For this touch their terminals a couple of times, as this will cause the potential difference between the two capacitor terminals to be neutralized. You may place a low value resistor across the leads to regulate the current flow rather than shorting the terminals.

VOLTMETERS

- Place the meter in the proper range of the expected voltage. If needed start at the highest range and then change the range to successively lower ones if the reading needs to be accurately determined. If the voltage to be measured is too high, and the meter is set in a much lower range of voltage measurement, the high voltage may damage the instrument.

- Connect the voltmeter across the element whose voltage is to be measured. Figure A-1 below shows the proper electrical connections for a voltmeter used to measure voltage across a circuit element.

AMMETERS

- Place the meter in the proper range of the expected current. If needed start at the highest range and then change the range to succes-

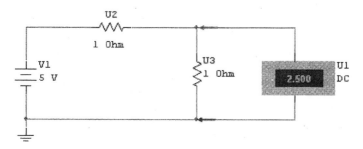

Figure A-1. Proper voltmeter connection

sively lower ones if the reading needs to be accurately determined. If the current flow to be measured is too high, and the meter is set in a much lower range of current measurement, the high current may damage the instrument.

- Event though most meters have high current protection fuses, it is best to first estimate the approximate range in which the measured value of current may lie, prior to use.

- Connect the voltmeter in series (inline) the circuit branch in which the current is to be measured. Figure A-2 shows both an Ammeter and a Voltmeter connected in a circuit.

DIGITAL MULTI-METERS

- Since these instruments can read voltage, current, resistance, etc., it is important that they be set to the proper parameter, prior to use.

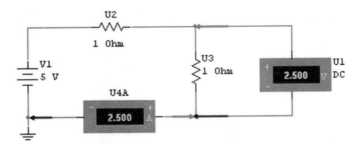

Figure A-2. Proper ammeter and voltmeter connections

- Most digital multi-meters are auto-ranging, meaning thereby that they can sense the proper range of values they are measuring.

- When using a multi-meter to measure voltage it should be used to measure across the element; and when used to measure current it should be placed in series with the branch in which the current flow is to be measured.

- When using the multi-meter to measure resistance ensure that the element is preferably not connected to any other circuit element. If the resistance must be measured without disconnecting the circuit, ensure there are no voltage sources (capacitors are discharged as well), and that at least one terminal of the resistor is not electrically connected to anything. This will ensure that only the resistance of the element will be measured by the multi-meter.

- Figure A-3 shows a representative digital multi-meter that is used to measure voltage, current and resistance.

LOGIC PROBES

A logic probe is a hand-held test instrument which detects the presence or absence of voltage at key test points. The operational status of a logic circuit or device can be evaluated with this device. As a rule, LEDs are used to identify the status of signals applied to the probe. In some probes, an illuminated green LED indicates the on or 1 status of a test point. An off LED or the illumination of an alternate red LED identifies the off or 0 state. Illumination of both LEDs indicates a bad level or open condition that is some-

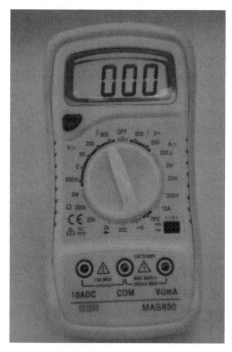

Figure A-3.Digital Multi-meter

where between logic 0 and 1.

Manufactured logic probes resemble an oversized pen with a pointed tip. This type of probe is generally energized by the circuit under test. Clips are attached to the + Vcc and ground of the circuit to energize the probe. The probe tip is then touched to a selected test point. If a voltage of approximately + 2.4 V is present, the 1 LED will light. If the voltage at the tip is 0.6 V or less, the 0 LED will light. Some logic probes have only one LED in their construction. This type of probe illuminates the LED with + 2.4 V and has the LED off when the voltage is less than + 2.4 V. Manufacturers of the probe generally provide instructions that identify these features.

Logic Pulser

The logic pulser is an ideal companion for the logic probe. Pulsers generate high or low logic levels or trains of pulses. This signal can be injected into a gate while observing the response of the output with a logic probe.

Logic pulsers receive operating energy from the circuit under test. When the probe tip is touched to an input point, a pulse is injected by depressing a pushbutton. An LED generally shows when the tip is 1 or 0. The LED is on with a 1 and off with a 0. A logic pulser is also called a digital signal injector. Pulsers usually have single-pulse or continuous-train generating capabilities.

MISCELLANEOUS EQUIPMENT

- Use the proper tools meant for specific purpose. Do not use a knife to function as a screwdriver.

- Make sure that all meters are calibrated.

Cables
- Use insulation tape over any electrical cables which are not insulated.

- Use cables rated for the proper voltage and current. For electronics use breadboard typically require AWG (American Wire Gauge) #22-24, solid wire.

- Do not use multi-strand wire with a breadboard as errant strands could find their way into neighboring slots and cause the circuit to malfunction.

FOR DEVICES

Static Electricity

- ESD (Electro Static Discharge) occurs when one material which is charged comes in contact with one which has a low charge or is grounded.

- Electronic components can be adversely affected by as little as 25 Volts even though humans will not mot sense this voltage.

- No polyester

- Use anti-static bags and anti-static straps, anti-static mats, especially while dealing with CMOS elements.

- Best environment to prevent static buildup is hot and somewhat humid.

- Worst environment for static buildup is cool and dry.

- Keep away from strong magnetic or electro-magnetic fields.

FOR ENTIRE LABORATORY

All activities should be done with low voltages whenever possible. Instructions should be written, with clear directions, for performing lab activities. All lab or shop work should emphasize safety. Experimental circuits should always be checked before they are plugged into a power source. Electrical lab projects should be constructed to provide maximum safety when used.

Disconnect electrical equipment from the source of power before working on it. When testing electronic equipment, such as TV sets or other 120-V devices, an isolation transformer should be used. This isolates the chassis ground from the ground of the equipment and eliminates the shock hazard when working with 120-V equipment.

A good first-aid kit should be in every electrical shop or lab. The phone number of an ambulance service or other available medical services should be in the lab or work area in case of emergency. Any accident should be reported immediately to the proper school officials. Teachers should be proficient in the treatment of minor cuts and bruises. They should also be able to apply artificial respiration. In case of electrical shock, when breathing stops, artificial respiration must be started immediately. Extreme care should be used in moving a shock victim from the circuit that caused the shock. An insulated material should be used so that someone else does not come in contact with the same voltage. It is not likely that a high-voltage shock will occur. However, students should know what to do in case of emergency.

Normally, the human body is not a good conductor of electricity. When wet skin comes in contact with an electrical conductor, the body is a better conductor. A slight shock from an electrical circuit should be a warning that something is wrong. Equipment that causes a shock should be checked immediately and repaired or replaced. Proper grounding is important in preventing shock.

Safety devices called ground-fault circuit interrupters (GFIs) are now used for bathroom and outdoor power receptacles. They have the potential of saving many lives by preventing shock. GFIs immediately cut off power if a shock occurs. The National Electrical Code® specifies where GFIs should be used.

Work surfaces in the shop or lab should be covered with a material that is non-conducting, and the floor of the lab or shop should also be non-conducting. Concrete floors should be covered with rubber tile or linoleum. A fire extinguisher that has a non-conducting agent should be placed in a convenient location. Extinguishers should be used with caution. Their use should be explained by the teacher.

Electrical circuits and equipment in the lab or shop should be plainly marked. Voltages at outlets require special plugs for each voltage. Several voltage values are ordinarily used with electrical lab work. Storage facilities for electrical supplies and equipment should be neatly kept. Neatness encourages safety and helps keep equipment in good condition. Tools and small equipment should be maintained in good condition and stored in a tool panel or marked storage area. Tools that have insulated handles should be used. Tools and equipment plugged into convenience outlets should be wired with three-wire cords and plugs. The purpose of the third wire is to prevent electrical shocks by grounding all metal parts connected to the outlet.

Soldering irons are often used in the electrical shop or lab. They can be a fire hazard. They should have a metal storage rack. Irons should be unplugged while not in use. Soldering irons can also cause burns if not used properly. Rosin-core solder should always be used in the electrical lab or shop.

Electricity causes many fires each year. Electrical wiring with too many appliances connected to a circuit overheats wires. Overheating may set fire to nearby combustible materials. Defective and worn equipment can allow electrical conductors to touch one another and cause a short circuit, which causes a blown fuse. It could also cause a spark or arc which might ignite insulation or other combustible materials or burn electrical wires.

ELECTRICAL FIRES

*   Fire extinguisher of type C is approved for electrical fires. Type A is used for wood, paper; while Type B is used for flammable liquids such as petroleum based products.

*   When using a fire extinguisher, aim at the base of the fire.

*   Always face the fire, and if necessary back away from it even while facing it. This will ensure that the user is aware if it suddenly spreads to a different location.

BREAKERS AND FUSES

*   Fuses should be of the proper rating for handling the rated load. Do not replace fuses with ones of higher rating as these might damage the equipment when rating is exceeded.

*   Fuses and circuit breakers are important safety devices. When a fuse blows, it means that something is wrong in the circuit. Causes of blown fuses could be:

— A short circuit caused by two wires touching
— Too much equipment on the same circuit
— Worn insulation allowing bare wires to touch grounded metal objects such as heat radiators or water pipes

- After correcting the problem a new fuse of proper size should be installed. Power should be turned off to replace a fuse. Never use a makeshift device in place of a new fuse of the correct size. This destroys the purpose of the fuse. Fuses are used to cut off the power and prevent overheated wires.

- Circuit breakers are now very common. Circuit breakers operate on spring tension. They can be turned on or off like wall switches. If a circuit breaker opens, something is wrong in the circuit. Locate and correct the cause and then reset the breaker.

- Always remember to use common sense whenever working with electrical equipment or circuits. Safe practices should be followed in the electrical lab or shop as well as in the home. Detailed safety information is available from the National Safety Council and other organizations. It is always wise to be safe.

POWER CONDITIONS

Being aware of the kinds of power problems which can occur can be useful in identifying what can be done to solve the problem.

Blackout
This occurs when there is a complete power failure, and no power is available

Brownout
This occurs when there is an under-voltage in the system which lasts for a short duration (few seconds).

Sag
This occurs when there is an under-voltage in the system which lasts for a very short duration (milliseconds).

Spike

This occurs when there is an over-voltage in the system which lasts for a very short duration (milliseconds).

Surge

This occurs when there is an over-voltage in the system which lasts for a short duration (seconds).

MAINTAINING POWER

Surge Suppressers/Protectors

These are useful against short term over voltages, and offer some protection for a limited number of such incidents. These devices often include an indicator which indicates whether it is functioning properly.

POWER CONDITIONERS

These are useful for monitoring the quality of power being received.

UPS (UNINTERRUPTIBLE POWER SUPPLY)

These supply power to the equipment they are connected to, by first converting the available AC power from the room wall output into DC then recreating the AC waveform. Owing to this conversion, line conditioning circuits are often incorporated into the UPS, to remove any electrical surges and to smoothen out the resulting AC. These devices have a VA (Voltage x Ampere) rating which should not be exceeded. A 500 VA rating for example means that equipment running on 110 Volts should not consume more than 4.75A. The UPS keeps functioning for a short while even after there is a blackout, giving users enough time to perform shutdown operations on their equipment.

LIGHTING

•   Adequate lighting is needed while performing all electronic wiring and testing.

- Adequate laboratory space is needed to reduce the possibility of accidents.

- Proper ventilation, heat, and light also provide a safe working environment

CIRCUIT CONNECTIONS

- Always use a circuit diagram while connecting up a circuit. Label all component values, mark voltage and ground connections.

- Keep the length of all wires in the circuit to a minimum.

- Use components well within their rated range. For example, a low resistance, low wattage resistor (say 10Ω, $1/4$ W) should not be used directly across with a 120V AC supply.

- Discharge capacitors prior to connecting them in a circuit.

- Take static precautions.

- Wiring in the electrical lab or shop should conform to specifications of the National Electrical Code® (NEC®). The NEC® governs all electrical wiring in buildings.

HUMIDITY, DUST, AND SMOKE

- Dust may clog equipment and cause heat buildup.

- Avoid direct sunlight for overheating purpose.

- Not too humid else there will be condensation on the equipment which could cause shorting owing to conductive path provided by the water.

- Not too dry, else there will be a possibility of static electricity.

- Humidity between 40-60% is recommended for electrical equipment.

HAZARDOUS MATERIAL

- Handling and disposal of any potentially hazardous equipment, cleaning agents, solvents, chemicals, etc., should be followed as per their MSDS (Material Safety Data Sheets).

Appendix B

Datasheets

Datasheets for the following digital ICs may be obtained from manufacturer websites such as On Semiconductor (http://onsemi.com), Fairchild Semiconductor (http://www.fairchildsemi.com/), National Semiconductor (http://www.national.com):

S.No.	Function	IC Number
1	AND gate	74LS08
2	OR gate	74LS32
3	NOT gate	74LS04
4	XOR gate	74LS86
5	XNOR gate	74LS266
6	NAND gate	74LS00
7	NOR gate	74LS02
8	NOT gate, Schmitt	74LS14
9	Comparator	74LS682
10	Monostable multivibrator	74LS122
11	Timer	LM555
12	Comparator (8 bit)	74LS682
13	Multiplexer: 1 of 4	74LS153
14	Multiplexer: 1 of 8	74LS151
15	Demultiplexer: 4 to 1	74LS156
16	Demultiplexer: 8 to 1	74LS156
17	Encoder (priority): 8 to 3	74LS148
18	Decoder: 2 to 4	74LS139
19	Decoder: 3 to 8	74LS138
20	Latch	74LS373
21	JK flip-flop, positive edge	74LS73A
22	JK flip-flop, negative edge	74LS76
23	D flip-flop	74LS74
24	D flip-flop, with tri-state o/p	74LS374
25	Binary counter: up/down	74LS169
26	Binary counter: synchronous	74LS160
27	Shift register	74LS95B
28	BCD to 7-segment display: common anode	74LS47
29	BCD to 7-segment display: common cathode	74LS48
30	Arithmetic and logic unit (ALU)	74LS381
31	Tri-state buffer, high Enable	74LS126
32	ADC	ADC0801
33	DAC	DAC0800

Appendix C

Constructing Digital Circuits

RESISTOR COLOR CODES

The value and tolerance of resistors are represented by four colored stripes. The first three stripes are bunched together and represent the actual value of the resistor, whereas the 4th stripe located a bit to the side of the other three denotes the tolerance. This is shown in Figure C-1(a) and the values of the stripes corresponding to different colors are shown in Figure C-1(b). The tolerance strips can be gold (5% tolerance) or silver (10% tolerance).

Example: Stripe color – Red, Black, Orange, Silver

$20 \times 1000 \pm 10\%$ Ω
$= 20000\ (\pm10\% \text{ of } 20000 = 10/100 \times 20000)$ Ω
$= 20000 \pm 2000$ Ω
$= 18000$ to 22000 Ω

		Orange		
=	Red Black × 10	± Silver%		
=	2 0 × 10^3	±10%		
≈	2 0 k Ω			

$20 \times 1000 \pm 10\%$ Ω
$= 2000\ (\pm10\% \text{ of } 20000 = 10/100 \times 20000)$Ω
$= 20000 + 2000$ Ω
$= 18000$ to 22000 Ω

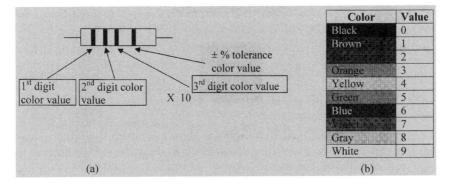

		Color	Value
		Black	0
		Brown	1
	± % tolerance	Red	2
	color value	Orange	3
1st digit color value	2nd digit color value	Yellow	4
X 10 3rd digit color value		Green	5
		Blue	6
		Violet	7
		Gray	8
(a)	(b)	White	9

Figure C-1. Resistor color codes and lookup-table for corresponding values

OHM'S LAW

An electrical circuit is given in Figure C-2(a). It has been found for experimental observations that as the voltage is increased in a circuit so does the current. When the voltage is decreased the current decreases as well. This is shown in Figure C-2.

This phenomenon is represented using Ohm's law:
$V \alpha I$ (i.e., Voltage, V, 'is proportional to' the Current, I)
$V = R \times I$

R is the factor which relates the applied voltage and the current in the circuit. It depends on the material used, the length, the cross-sectional area and the temperature.

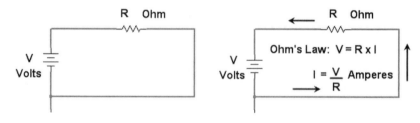

Figure C-2. Ohm's law

SAMPLE ELECTRICAL CIRCUIT

A simple electrical circuit is shown in Figure C-3. It consists of a 5V voltage source, a switch and a 25 Ω resistor. If the switch is open no current can flow in the circuit, and if the switch is closed current can flow.

INPUT/OUTPUT CONDITIONING

In general the inputs to a digital system are not in the range which may be directly sensed as either a low-level or a high-level by a digital circuit. For example the receiver circuitry recording information from a distant undersea or space vehicle may receive voltages which are very low in amplitude, so much so that they may never cross the high-range value

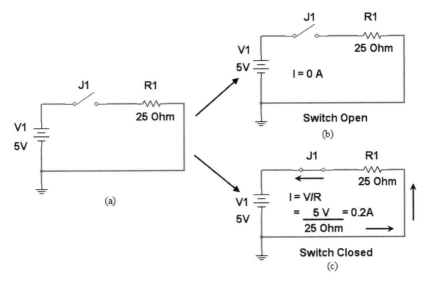

Figure C-3. Operation of a electrical circuit

required. In other cases the voltages in use by say an industrial system, typically rated at 120V AC may be too high to serve as inputs to a digital system. In such cases the signals need to be transformed into values which correspond to the low- and the high-range.

On the other hand where a digital output is concerned, this is typically in the range of 0V – 5V, with only a small amount of current capability in either the low or the high range. Thus it will not be able to drive any electrical equipment which has even moderately high current requirements such buzzers or any AC equipment such as pumps or industrial motors. Thus the final outputs generated by a digital system need to be conditioned with regards to voltage, current or power requirements.

Conditioning Inputs to Digital Circuits

Examine the block diagram of a digital system shown in Figure C-4.

Assuming that the digital system is built using TTL technology, all the devices are powered with 5V and have a ground (0V) connection. Further when a signal with amplitude in the range 2V – 5V is applied to the input of any of these devices it is sensed as a high; while if it is in the

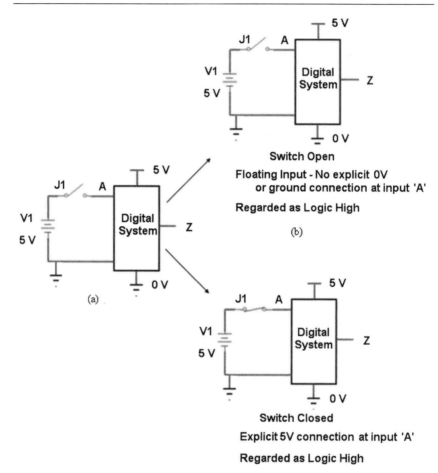

Figure C-4. Illustration of 'floating-high' in a digital circuit built using TTL gates

range 0V – 0.8V, it is sensed as a low.

Now we are going to examine the behavior of the system when the switch J1 is closed and when it is open. On closing the switch 5V is applied to the digital system. However, when the switch J1 is open, no specified voltage is applied to the digital system. It would seem that this corresponds to a 0V input, but in fact this is not so. The TTL devices used to create digital circuits are provided with power and ground connections. In the absence of an input, such as when the switch is 'open', since this does not explicitly provide a voltage in the range 0V – 0.8V, these devices will in most cases treat the 'open' input as a High. So instead of a Low, a High is sensed by the digital system. Thus, regardless of whether the switch is open or closed the input read in is a High.

It should be noted that the ground (0V) of the digital system should be the same as that used by the ground for the circuit providing inputs to it. With the same ground reference for the input, processing and output parts of the digital system the voltage at any point can be determined consistently.

It is important not to leave any of the inputs on the digital IC being used which could potentially affect the output of the device 'floating'(not explicitly connected to anything). Moreover, such floating inputs are liable to pickup external electrical noise in the form of voltage signals, and this can lead to erratic output from the chip.

Several arrangements of resistors and voltage sources have been created such that they provide explicit 5V for High and 0V for Low as inputs to the digital system.

PULL-UP RESISTOR ARRANGEMENT

When the switch is open (off) the voltage applied to the digital IC input gets pulled up to approximately 5V. Common resistor values for this arrangement range between 1kΩ – 10kΩ. Typically a 4.7kΩ will suffice. Figure C-5 shows this operation.

PULL-DOWN RESISTOR ARRANGEMENT.

When the switch is open (off) the voltage applied to the digital IC input gets pulled down to approximately 0V. Common resistor values for

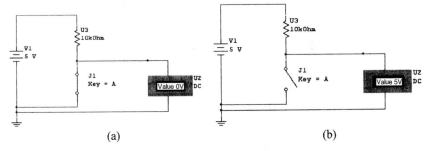

(a)   (b)

Figure C-5. Operation of a pull-up resistor arrangement

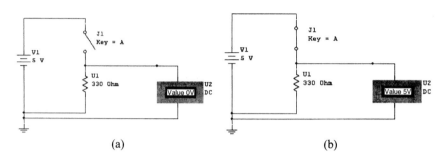

(a)   (b)

Figure C-6. Operation of a pull-down resistor arrangement

this arrangement range between 100Ω – 1kΩ. Typically a 330Ω will suffice. Figure C-6 Shows this configuration.

CONDITIONING DIGITAL CIRCUIT OUTPUTS

Light Emitting Diode

Light Emitting Diodes or LEDs belong to a class of electronic devices called diodes. These devices emit light when an appropriate voltage source is connected across their terminals and sufficient current flows through them. They are used as indicators, in displays, and are steadily gaining in popularity. The circuit symbol and the sketch of an LED is shown in Figure C-7.

DC voltage sources have terminals marked + and – . Electrical lamps on the other hand do not have any specific polarity marked on them, and

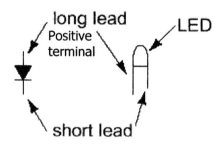

Figure C-7. Circuit symbol and sketch of an LED

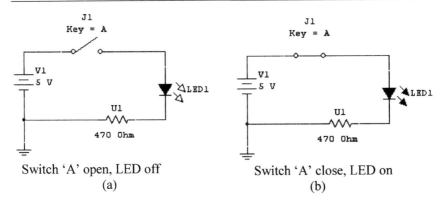

Switch 'A' open, LED off   Switch 'A' close, LED on
(a)   (b)

Figure C-8. Operation of an LED

regardless of the manner in which the lamps terminals are connected to a DC voltage source, it will glow. However LEDs do have a polarity as designated in Figure C-7.

LEDs emit light only when they are provided with the proper polarity voltage and current. The positive terminal of the supply needs to be connected to the positive terminal of the LED, and the negative terminal of the supply to the negative terminal. Such a configuration is called forward biasing the LED. This is shown in Figure C-8.

When using a 5V DC source, and an LED, it is important to limit the current which is flowing in the LED circuit to less than 25mA (25×10^{-3} A). Refer to datasheets for the LED being used in order to determine the largest amount of current which can flow through it. LEDs themselves use around 2V for their operation, and do not offer any significant resistance. The remaining voltage 5V-2V = 3V is to be dropped down to zero by passing it through a resistor. This 'current limiting resistor' should be such that it will limit the current to 25mA in the overall circuit.

Using Ohm's law:

$V = I \times R$

$3 = 25 \times 10^{-3} \times R$

$3 / (25 \times 10^{-3}) = 400 \ \Omega$

Typical values of the current limiting resistor range between 330 Ω – 1000 Ω. If the value of the resistor is too small, an excessive amount of current will flow in the circuit, overheating the LED and reducing its useful life. On the other hand, if the value of the resistor is too large, only a small amount of current will flow, and the LED will be dim or may not switch on at all.

7-SEGMENT DISPLAYS

The 7-segment display shown in Figure C-9 is very commonly used in digital systems today.

Figure C-9. 7-segment display

These electronic devices are usually designed to display a single decimal character. Light emitting diodes (LEDs) can be conveniently packaged into 7-segment displays for indicating alphabets and numerals. LEDs can be made to emit light by providing the proper polarity (positive to anode terminal 'A', the longer leg, and negative to cathode terminal 'K', the shorter leg), while ensuring that the current flowing through the device does not become too high.

When the negative terminals (cathode) of all the LEDs used in the package are shorted together, it is termed as a common-cathode display. In order to light up a particular LED or group of LEDs in the display, appropriate voltage has to be applied across the LEDs and adequate current should flow through them. The operation of the display is shown in Figure C-10 by switching on only one of the 7 segments using a switch.

DRIVING HIGH POWER DEVICES

Designing a working digital circuit to control a mechanism requires a working knowledge of how the overall system is supposed to function (including the mechanical, pneumatic, hydraulic, etc devices it contains),

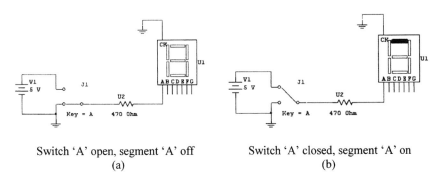

Switch 'A' open, segment 'A' off   Switch 'A' closed, segment 'A' on
(a)   (b)

Figure C-10. Operation of a 7-segment display

a clear understanding of the control objective, the inputs connected to the system, and the outputs it is required to switch on or off. After the digital circuit is built it is interfaced with input sensors and possibly high-power outputs as well.

It should be noted that many industrial and office equipment require substantially higher voltages than the 5V DC which is the standard output of the TTL logic chips used to design digital circuitry. Suggest or sketch how higher voltage, amperage or power device can be switched on and off using the 5V DC output from TTL chips.

A common design strategy while designing a digital circuit is to separate the control part of the circuit (the design done using logic gates) from the power part of the circuit. Then these parts can be independently worked on, or even replaced entirely without affecting the other. With this in mind, we will be using the output of the digital circuit to switch a transistor on and off. The transistor which has three terminals base, emitter and collector essentially functions as an electronic switch.

The base of the transistor is responsible for switching it on and off. When a voltage is applied to the base of the transistor (with the collector, emitter have been connected to appropriate voltage and ground), then the transistor switches 'on', acting as a 'closed or shorted' switch. Doing so will connect the collector and emitter terminals. Any device connected in the collector-emitter circuit at that time will experience a flow of current through it, and will turn on itself. Place a small lamp or motor in the emitter-collector circuit. Using this simple scheme a low voltage (0-5V output) from a digital circuit can be connected to the base, whereas a high power appliance (lamp, motor) can be connected in the collector-emitter circuit. This is shown in Figure C-11.

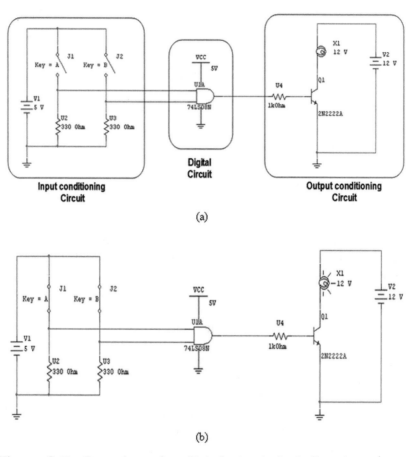

(a)

(b)

Figure C-11. Operation of a digital circuit including input/output conditioning

Index